MOBILE USABILITY

MOBILE USABILITY

How Nokia Changed the Face of the Mobile Phone

Christian Lindholm
Nokia
Helsinki, Finland

Turkka Keinonen
University of Art and Design Helsinki
Helsinki, Finland

Harri Kiljander
Nokia
Helsinki, Finland

McGraw-Hill
New York   Chicago   San Francisco   Lisbon   London   Madrid
Mexico City   Milan   New Delhi   San Juan   Seoul
Singapore   Sydney   Toronto

The McGraw·Hill Companies

Library of Congress Cataloging-in-Publication Data

Mobile usability : how Nokia changed the face of the mobile phone /
   Christian Lindholm, Turkka Keinonen, Harri Kiljander [editors].
       p.   cm.
   Includes bibliographical references and index.
   ISBN 0-07-138514-2 (alk. paper)
       1. Cellular telephones.   2. Nokia (Firm)   3. User interfaces (Computer
   systems)   I. Lindholm, Christian, date   II. Keinonen, Turkka.   III. Harri
   Kiljander.

   TK6570.M6.M597   2003
   384.5'35—dc21
                                                              2002044394

*The sponsoring editors for this book were Marjorie Spencer and Stephen S.
Chapman, the editing supervisor was David E. Fogarty, and the production
supervisor was Sherri Souffrance. It was set in Candida by North Market
Street Graphics.*

*Illustrated by Valtteri and Cleo Bade. Photography by Tapani Pelttari, authors,
Nokia image bank, and NASA Archives.*

Printed and bound by RR Donnelley.

 This book was printed on recycled, acid-free paper containing a minimum
of 50% recycled, de-inked fiber.

McGraw-Hill books are available at special quantity discounts to use as
premiums and sales promotions, or for use in corporate training programs. For
more information, please write to the Director of Special Sales, Professional
Publishing, McGraw-Hill, Two Penn Plaza, New York, NY 10121-2298. Or
contact your local bookstore.

CONTENTS

PREFACE

This book is different from many others about user interface design and usability in that all chapters are about one company—Nokia—and all are written by Nokia employees, present or former. In other companies, books about user interface (UI) design, such as those by Bergman[1] and Wiklund,[2] are typically written by one or two authors presenting a coherent opinion. The latter gives readers a harmonious idea of how things work there, what represents state of art in the company, and what needs to be improved. Reading such material from a Nokia point of view, especially for the editors of this book, creates ambivalence. *How can they keep the whole thing on track so well? Do they actually reach agreement on everything?* Sure, they have the same problems with people outside the usability or design organization who don't "get it," but inside the organization communications seem clear and perspectives aligned. What happens when team members have something else to say?

The picture of user interface design and usability research that readers will take away from this book is admittedly more fragmented and contentious. This outcome is partly planned, given that we set out to present a whole range of usability approaches at work, but partly it took the editors by surprise. We had asked authors to emphasize their personal conclusions about what is essential and what they themselves learned in their projects. They did that. The result was a collection of experiences and opinions that do not seem to fit together. What is the Nokia usability process? Which methods do we use? How are users integrated into design work? How is design integrated with research? Who makes the decisions? What resources are needed? What is the essence of mobility in user interface design? Can we stay in the lab, or do we need field tests? What is our vision? What is the core of our expertise? The answers depend on whom you ask. Authors with different responsibilities and different professional backgrounds work in many positions throughout the organization. They care about different things and raise different issues. Their opinions vary accordingly.

Is there anything common to all the chapters? Let me answer by trying out an analogy on you. The typical household may be completely satisfied with its compact camera. Every feature in the camera is consistent with the whole, making the object easy to handle. There are no conflicts between the parts of the product. It is cheap to buy, and the quality of the pictures is good—when you snap in favorable circumstances with a lot of light, that is, and if your subject stands still and you get close enough. At least the pictures are good enough for the family album. The camera does the job, and this is the job it was built to do.

The authors in this book don't get to talk about well-lit stationary subjects in close range. They have to talk about elusive subjects that are difficult to expose and that demand specialized solutions. At the end of the day the image captured will be cleaned, sized, and sharpened for printing on the cover of the most widely read magazine in the world, *The Public Record.* No camera has been built for this job; the whole bag of equipment is needed.

Everything begins with a reliable instrument. The best film, the best lenses—your camera has to be flexible enough to accept them. All your valuable accessories and your special skills—they'd be of little use if you needed a different camera body for each shot. The chapters in Part 1 reveal the framework of Nokia user interfaces. User interface styles provide the

A reliable and robust instrument.

common logic for all new features, and a majority of user interface creation activities culminate in solutions matching existing styles. Styles have a huge influence on the user's experience with our products, and internally they streamline software processes in implementing new models.

They also are everyday tools used by the designers. *Can you realize that feature with Navi-key style?* The styles are surprisingly flexible; more and more new features but very few new styles have been added. Once in a while new features do challenge the limits of existing styles, but not often. Harri Kiljander and Johanna Järnström (Chap. 1) introduce Nokia user interface styles and discuss the principles of maintaining an appropriate style portfolio. Seppo Helle, Johanna Järnström, and Topi Koskinen (Chap. 2) focus on style elements: the mobile menu, control keys, and user interface graphics. While keys and their functionality remain pretty consistent with a styles approach, the part of a mobile UI that bears the brunt of accommodating new features is the menu.

The right lenses, reflectors, and filters for your camera are determined by your subject, your style, and the purpose of the picture. With a standard 50-mm usability lens something essential may be cropped out. You can, of course, step back, or in our case wait until you see the effects of a design solution once implemented. But if you need to include more of the scenery from where you stand right now, you need a contextual wide

Sociological wide-angle lens—FD 28 mm 1:2.8 for contextual user comprehension.

angle of 24 mm, maybe even a sociological fisheye. With these lenses, everything that is important will be exposed. You will get the big picture, with things in their proper places. The details may not be particularly sharp, but they are there to be seen. In Part 2 Katja Konkka (Chap. 4) gives us a panorama picture from Mombai (formerly Bombay), India. The richness of detail is amazing. Riitta Nieminen-Sundell and Kaisa Väänänen-Vainio-Mattila (Chap. 5) explain the advantages of the fisheye and teach us how to zoom out.

Suppose that you glimpse something that looks interesting on the horizon, but can't quite make it out. You start digging in your camera bag for a telephoto lens. With that you can focus on telling details that are hard to discern from a distance. Using a 400-mm telephoto vision lens calls for special skills, of course. You can easily miss a moving target, and it may be difficult to focus the lens precisely and keep it steady enough for the instant of exposure. Maybe you'll need ultraviolet filters to reveal objects in the haze. And finally you have to load the camera with extremely sensitive film to compensate for all the complicating factors. Panu Korhonen and Pete Dixon (Chap. 12) try these tricky gadgets. Do they have an image stabilizer?

When you start to be pretty sure that the object in your crosshairs is what you're looking for, the next step is to set up a tripod. With your camera and telephoto on a firm footing, the image will be much clearer. The firm basis on which user interfaces are built is the enabling communication technology. Jussi Pekkarinen and Jukka Salo (Chap. 10) talk about their efforts to erect a platform from which to make a good picture of the mobile Internet. In their story the target was in sight, but freehand images were never sharp enough.

Any professional photographer's bag includes a selection of lenses for proper granularity. Miika Silfverberg (Chap. 7) writes about the macro lenses and extension tubes with which he focuses on the users' ability to input text. The setting for such a picture must be carefully constructed. You'll need the right kind of lights positioned at the perfect angles; you'll need time to check the aperture. Perhaps you'll want to consult the exposure tables and consider correction factors. To control all the variables, it's best to stay in the studio. Often this technology reveals details that are not visible to the naked eye, even the most experienced one.

Lab test extension tubes—extension tube set M20 for optimizing critical details.

There are always times when a photographer is just not sure about composition or lighting, even with all the right equipment deployed. Maybe she has a fair idea of what will happen when she depresses the shutter on the basis of her professional experience, but the idea never matches the real image perfectly. She could shoot each picture and

Iterative instant camera—Polaroid for design iterations and discussion.

process and print it on site, but that takes too much time at a live shoot. Her solution is to snap a Polaroid, wait the few seconds for the test image to appear, maybe show it to the clients, adjust the composition as necessary, and snap again. Only when everything meets her standards does she load the camera with real film. Ketola (Chap. 8) and Keinonen (Chap. 9) also use Polaroid images. They have a normal 50-mm usability lens, but their film is selected for adjusting angles, not for taking the picture.

On the street, all your options are in your bag. Some photographers are continually on the move. They never know whether their next assignment will be breaking news, an environmental portrait, a football match, or an annual report. Carrying enough equipment for all contingencies would be backbreaking, so they get practical. A zoom lens may not give ideal exposures, but it replaces a whole lot of special-purpose glass. Lindholm's managerial approach to user interface development (Chaps. 3 and 6) requires us to zoom in and out on cue. In an acceptable result, the details have to be recognizable and we must be able to tell where they belong.

John Rieman (Chap. 11) writes about people who shoot action. They do not have time to take meter readings or Polaroids; they need to act in the moment. If you start to doubt, the moment is lost. These guys need to master their camera and the all-purpose 50-mm usability lens without any

Infinite zoom from telephoto to microscope in one turn for design management vision to design management attention to detail.

Heuristic flash—740-1 for
decision heuristics.

hesitation. A flash gives them the only tool they need to shoot in any circumstance.

The camera bag metaphor is misleading in several aspects, and inspired by technologies that belong to the past. Still, I hope it helps convey the message that for an enterprise delving deep into user interface design and usability, and an enterprise counting on the expertise of individuals, objects of study present themselves through the lens of perspectives and contexts, not as formal entities. There start to be more shades, more approaches and more opinions. The approaches are neither right nor wrong—they are different.

References

1. E. Bergman, ed., *Information Appliances and Beyond, Interaction Design for Consumer Products.* Orlando, Fla.: Morgan-Kaufmann, 1999.
2. M. E. Wiklund, ed., *Usability in Practice: How Companies Develop User Friendly Products.* Boston: Academic Press, 1994.

ACKNOWLEDGMENTS

New technologies leading to third-generation mobiles gradually widen the scope of mobile services and mobility in general. As markets grow they increasingly get segmented, and new opportunities arise with them. Mobile terminals have changed to open platforms for third-party software developers. In the 1990s the people directly involved in designing mobile user interfaces were very few. Now mobile services and downloadable applications start to enable much wider participation in the creation of mobile user experiences. It was time to address the people doing pioneering work in open mobile platforms by leveraging some of the lessons we have learned from designing different mobile UIs during the past decade. We wanted to give context to all the development tools and material we are publishing. This book is for you; let your creativity flow free.

Nokia has been instrumental in creating benchmark mobile user interfaces during 1990s. Summarizing this work and spreading our experiences to a larger audience has, we hope, a potential for contributing to the work of all who attempt to design usable, interesting, and perhaps delightful mobile experiences.

The user interface design culture, which this book describes, first came to life through the influence of a few mobile interaction design pioneers in the company. It rapidly matured as an increasing number of specialists joined the team. These people, who are much too numerous to be listed here, have created the designs and methods described in this book. We want to thank you for everything we have learned and for the enjoyment in work you have given us.

In the beginning of 2001 we asked for contributions to a book on Nokia user interface design and usability from our Nokia colleagues. We received many interesting and personal proposals, many of which we had to reject due to the practical limitations of the book. We want to acknowledge all the contributors who spent time and effort in bringing this story public.

During the process of editing the book Turkka Keinonen moved from the Nokia Research Center to the University of Art and Design Helsinki.

We want to acknowledge the Department of Product and Strategic Design for allowing Turkka to finalize the project. Harri Kiljander became Director of User Interfaces at Nokia Mobile Phones in the beginning of 2003 and Christian has taken on new challenges in Nokia.

In addition, we want to acknowledge Nina Lindholm, the wife of Christian, and Maarit Laanti, the wife of Harri, for their loyal support. We are grateful to Valtteri and Cleo Bade for visualizing the concepts of the book, Tapani Pelttari for his creative use of photography, and Tapio Hedman and Eivor Biese for supporting the book in its early stages.

Finally we want to thank all those Nokia users who have personally provided us with insightful opinions and inspiration.

Christian Lindholm
Turkka Keinonen
Harri Kiljander
Helsinki

Turkka Keinonen

Introduction
Mobile Distinctions

Developers and designers have been taught that people are not crazy about new technologies but they appreciate real benefits in products. At the same time, we have set up our booth in a marketplace where this understanding is frequently blurred by a haze of enthusiasm cast up by emerging technology.

We, too, contribute to the ascendency of technology. Being in the vanguard of early adopters is a real "must," we say. Being first to find and use a new solution is a value in itself. These arguments are stated overtly and even proudly. Behavior that can be understood only on the basis of such motives is common, and can often appear in the guise of fulfilling user needs. *We all need to be able to control our coffeemakers with our handsets!* Why? Because it is possible. Because if you don't do it now, someone else will do it before you.

The enthusiasm is understandable and probably even unavoidable. Technology raises passions. Is there anything more rewarding than making something work for the first time when you know it is doable, but no one has done it yet? Making it faster than it has ever been? Making it just a little bit cleverer? For a big proportion of R&D activities, new and sophisticated technologies are the goal. There is nothing wrong in that. It is not necessary to demand that the everyday needs of nursery school teachers or ironworkers be considered in the early development phases of low-power radio-frequency transmissions. However, it is vital that user needs meet the technologies at *some* point in the development cycle, or they will die. The technologies, that is. The customers will survive nicely without the latest gadgets.

Our experience has been that the intersection of user needs and the industry interests increasingly takes place only *after* product launch. The birth of a mobile communication culture has been so sudden that it's no wonder that unpredictable phenomena occur. New solutions are utilized in ways that never even occurred to their designers. That is acceptable, of course, but designers would feel a lot better if given a chance to take user behavior into account before the launch. This book is about arranging a summit—the heads of the superpowers, *industry* and the *consumer,* meet in a user interface.

It won't be the first time they've met. Why is this case of special interest? We think that industry and the user will find it exceptionally challenging and worthwhile to agree on mobile communication technology. The reasons for this belief are rooted in mobility, communication, technology, and design.

Sequential Presentation

Compared to desktop solutions, user interfaces in mobile phones are pretty simple. There are far fewer features, even though the number is rapidly increasing. User opportunities to manipulate user interface (UI) objects are essentially more limited than with PCs. With only a couple of keys, the user can hardly select a wrong one. Where is the challenge? Aren't the problems just a fraction of those with desktop interfaces, a small subtask for a desktop application designer? Is mobility an excuse for us to distinguish our position among the mainstream desktop user interface design experts?

Mobile phones get carried around. For some extreme users, terminals follow wherever they go. For the rest, terminals are at hand for a significant proportion of their active time. Phones reside in pockets, in handbags, on belt clips, and in holsters. If these devices did not accompany the user, the main benefit of a mobile phone would be lost—it wouldn't provide immediate access anywhere any more, it wouldn't render the user reachable, it wouldn't be so personal. It would, in short, forfeit its bid to be our primary personal communication link to other people and to services.

The most obvious engineering solution to ensure portability is compact size. Smaller terminals allow more freedom in the way they are worn. The negative impact of decreasing size is also obvious—it influences the sizes of physical user interface elements. The total surface area of the terminal limits the number and size of buttons. The same applies to the screen. Given the limits of human vision, the amount of data that can be presented on the screen at a given time is very limited.

The majority of UI design and usability experience as a whole has been gathered from desktop interfaces, which are characterized by *direct manipulation.* Direct manipulation interfaces provide permanent representation of objects of interest, the physical operations that can be performed on those objects, and immediate visual responses to such manipulations, as displayed on those same objects. Desktops have Web interfaces with numerous options labeled by texts, most of them long enough to resemble natural language, or by icons, thumbnail images organized—when well designed—into visual structures that link relevant topics to support the user's search strategies. The difference in designing UIs for desktop environments versus phones is about quantity; desktops can accommodate more. However, when we delve a little deeper, the difference in quantity turns into a difference in quality. The logic of PC interfaces does not scale downward. Small interfaces are essentially different from big ones.

Parallel representation, in which plenty of options are displayed simultaneously on a sizable screen, turns into sequential representation on a small screen. You have to browse through options one by one because you can see at best a few at a time. We know that on a 17-in screen the user is able to navigate dozens of interactive items. Learning to use these interfaces fluently depends on the user's ability to match locations on the screen with related features. There can be a lot of mere stuff on the screen, but fluent users are easily able to pick out the ones they use most often from the rest. What is essential about parallel representation is that I may use different items for my purposes than you, and yet the presentation allows both of us to develop personal skills to match personal needs. For you and me, then, the UI becomes two different interfaces because we have learned to use it in different ways. Still the interface is the same from the manufacturer's point of view. The manufacturer's responsibility is to provide a set of features, show them all, and let the big screen and the user select the priorities.

Here I admit to exaggeration. On a desktop UI features are prioritized, albeit not very heavily. The user may, and often does, personalize the interface tools, but is able to develop personal styles of use even without customization of any sort.

A mobile user interface presents just a few items at a time. Users can't necessarily hit their selection right away, but must expect to browse the menu in the order that selections are presented. Everybody browses the menu in the same order. An easy way for users to opt out would be to avail themselves of shortcuts to their preferred items, or to tailor their menus as they wish. However, providing a means of tailoring does not remove our responsibility to provide usable defaults. Mobile phones are consumer goods, and we cannot expect the user to configure them. So, aside from knowing what kind of features to provide, we need to know—for each object in each situation—the most likely next few functions. It's easy enough to provide immediate access to a couple of options, but any user who is repeatedly forced to browse through, click by click, to the seven-teenth most likely selection will be beyond frustrated, whereas with a desktop interface with 40 icons in plain view, this user would never even notice that the first choice was in position 17.

Once we focus in on user needs, knowing the user allows us to select the most relevant features for most situations, which in turn enables us to present compact and portable designs. (See Fig. I.1.) It also enables us to provide cost-optimized solutions for people who don't need all the bells and whistles. Mobile sequential interfaces are fragile to bad design. We cannot provide all the options without reference to the essential ones. Mobility via limited output capabilities and sequential presentation leads inexorably to the requirement for feature prioritization.

Experts in Communication

Take a VCR, for instance. Someone rents a movie, sticks it into the player, hits a button—switches off brain, sits on couch, stays home, gets isolated and alienated. In that abbreviated tale there are elements of the machine, the service, and the society. Or take a microwave—you won't need to adjust the scenario all that much. The familiar buttons are on the panel,

Figure I.1 *The path of reasoning from portability to user understanding.*

and the pizza delivery service is a phone call away. Within the phenomenon of warming up prepared food may lurk culturally significant meanings: perhaps a demarcation line between men's and women's household responsibilities (she cooks food, he heats it) or perhaps an assumption about what kids of a certain age can be trusted to do.

With communication products the overlapping interaction layers of machine, service, and society are, we claim, more salient, transparent, and integrated than with most other products or technologies. Is this an appropriate hour to call someone? If I call today, am I being too pushy? Should I wait for her to call me instead? Maybe it would be better to send an email. Should I call her friend first? If I delete her number from my overpacked contact directory, am I abandoning her?

Communication—that motley mix of approaching others, getting to the point, wrapping it nicely with a suitable amount of small talk, giving a blunt response, hinting at your meaning without spelling it out, or pointedly ignoring something that was said or should have been—is an area where human skills are very advanced and very much culturally colored. There are strong conventions concerning what is acceptable for subordinates in an office to say, or for children to do. Men communicating to women, women to men, people in my trusted primary groups, total strangers—for all these situations, there are behavioral codes. To be accepted as a means of communications, technology must adapt to those codes. On the other hand, it appears that technology itself influences the codes and changes them.

The complexity of these phenomena makes task-oriented design difficult. Can you task-analyze social bonding? Taping a championship match with a VCR is a task that can be isolated and analyzed, and so is nuking a slice of pizza; but apologizing, negotiating, and reassuring are not so easily reducible to tasks. They are design challenges that call for core usability skills, debate on the social level of interaction, and understanding of the user's contextual and cultural position relative to that person's interlocutor.

The Hidden Possibilities of New Technology

Mobile phones used to be functionally direct replacements of their wired forebears. Now they have suddenly become platforms for entertainment and commerce and tools for information management and media consumption. One device starts to speak to another device without human intervention. One terminal supports protocols that other terminals don't understand. Successive product and service generations start to coexist. It is difficult to imagine all the different applications of these permutations. It is hard to predict which solutions will dominate and which will fail outright.

The mobile user experience will be influenced by user interface, the application in the terminal, the operator's solutions, the service provider's way of presenting service content, and the quality of the content itself. All these factors utilize various technologies, and they all have interfaces to one another. Creating even a simple service calls for cooperation between many experts and disciplines. In a service development project today there may be no one who understands the whole range of solutions. Often, UI designers with impressive backgrounds in human sciences or design have the most trouble following technical discussions. For the majority of project participants, therefore, the only way to get a working assumption of what the technologies enable us to do and how they are likely to be used is to be involved in these projects long enough. Even then, educated guesses and developed intuitions are only approximate. Something that was supposed to be easy to implement turns out to be practically impossible. Sometimes the opposite occurs. Solutions that were originally postponed to allow technology to catch up are suddenly realized in unexpected ways.

What are the consequences of all this? Not only are the possibilities of future technologies unpredictable but, on the level of individual projects, even the possibilities of present technologies may remain mysterious. Technology does not provide bedrock for design.

PART 1

Dominant Design in Mobile User Interfaces

In the relatively brief lifetime of the mobile phone, two major technological trends have taken hold: (1) devices trend toward miniaturization and (2) applications, features, and functions trend toward expansion. Taken together, they present us with an interesting paradox—squeezing more and more applications into smaller and smaller terminals is how we try to keep users satisfied, but it makes the devices harder to use. This apparent and obvious challenge has a deep impact on mobile user interface (UI) development and characterizes the user interface solutions introduced in the following chapters.

Technology is not a barrier in this case. Advances in miniaturization allow us to make radio transmitters and other components smaller than before. Processing power increases even as energy consumption decreases, so batteries also can be made smaller. Adding new applications is largely a question of imagination and software effort. Users want their phones smaller and lighter for ultimate portability, so all the new applications need to be managed with fewer, or smaller, buttons and displays—no technological barriers there. Still, human sight is not getting better to read smaller fonts; fingers are not getting thinner to press smaller buttons.

Figure 1 illustrates the progress of mobile phone technology since the early 1990s. All these phones are handheld GSM terminals. Nokia 1011 was the first GSM hand portable; the others are subsequent classic category phones.

The range of user needs grows every month. We are no longer serving business users alone. Games and other entertainment features are the most popular applications among young users, whereas to-do lists and calendars are more widely used by adults. Some users love to personalize their phones by downloading ringing tones and graphics, while others could not

MODEL	Nokia 1011	Nokia 2110	Nokia 6110	Nokia 6210	Nokia 6610
YEAR INTRODUCED	1992	1994	1997	2000	2002
DISPLAY TYPE	2 x 8 chars	3 x 10 chars + 2 x 6 chars	84 x 48 pixels	96 x 60 pixels	128 x 128 pixels
NUMBER OF KEYS	22	23	21	21	23
NUMBER OF DISPLAY TEXTS	406	378	1719	2777	3085
DEVICE VOLUME, CUBIC CENTIMETERS	340	170	130	95	71
DEVICE WEIGHT, GRAMS	475	240	140	114	84

Figure 1. The number of display texts reflects the number of software features. This figure is given per language, and it does not contain help texts. Since Nokia 6110, the software has also contained graphics, and individual texts are generally longer in the later models, so the differences in actual amount of software content are in fact somewhat greater than what numbers alone would suggest. Weights and volumes assume standard battery power.

care less. All kinds of wireless services are emerging, offering something for ever-widening populations of users. Although nobody would buy a phone without the basic phone applications like phonebook and text messaging, it's a rare consumer who won't want some of the numerous "indispensable" features offered by certain models. For product appeal, the longer the product's list of extras, the better, just in case. This mindset calls for utmost flexibility in the UI to accommodate new features in a mobile terminal with no accommodation for new special keys.*

Part I of this book addresses the way Nokia has chosen to approach the "squeezing dilemma" and other mobile challenges during the development of second generation mobile phones in the 1990s. We approach it from three angles. In Chap. 1, Kiljander and Järnström introduce Nokia's user interface portfolio and principles for developing and managing the user interface styles. In Chap. 2, Helle, Järnström, and Koskinen discuss the most comprehensive UI solution in Nokia mobile phones, namely, the menu—they explain its construction and the limits of menu-based user interfaces. Finally, Lindholm recounts the history of the Navi-key user interface style that, since its introduction in 1996, has been the most widespread mobile user interface in the world.

---

*The trend toward excessive functionality much beyond the actual needs of the users has been questioned repeatedly. Our statements here are not meant to imply an opinion about the benefits and drawbacks—business or human—of this trend, even though we recognize that we are stakeholders in the system. Our comments here are intended only to acknowledge the development that has taken place so far.

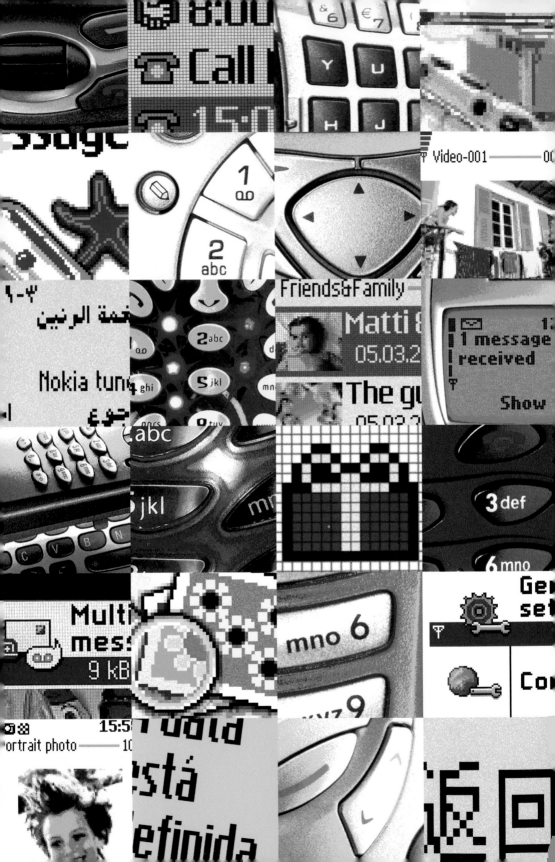

Harri Kiljander and Johanna Järnström

CHAPTER 1

User Interface Styles

An estimated 423 million mobile phones were sold globally in the year 2002. The mobile phone has become the information appliance that keeps people in touch. Among phone manufacturers, user interface is one of the key elements in the fight for customers. To the individual who purchases a mobile phone, a good user interface means convenience and an intuitive and easy access to the most frequently needed calling functions. It creates positive word of mouth and encourages brand loyalty when it's time for the customer to purchase a new phone. To the mobile operator, a good user interface leads to increased usage of the operator's wireless services and to reduced need for customer support.

Unlike the personal computer industry, the cellular industry has no standard user interface. Practically all mobile handset manufacturers are using their own UI solutions and conventions. This freedom offers wider choices to customers, lets manufacturers gradually develop and improve the user experience, and in general makes the user interface a competitive asset in the race for market dominance.

Usability and user-centered design are considered core competencies at Nokia. When we're talking about usable mobile phone user interfaces, the key artifact we want to keep in focus is the *user interface style*. (Some experts in the human–computer interaction field use the term *interaction style*). By this we mean the UI design conventions applied throughout a handset. This chapter describes the concept of UI style, briefly explains how to align customer segments with the appropriate user interfaces, discusses how user interface styles are created and evolved, and ponders the ways that the cellular mobile telephone user interface may change in the near future. Details of the UI styles themselves, namely, screen layouts, control key combinations, and menu structures, are discussed in Chap. 2.

Segments and Categories

Customers in Japan differ from customers in the United States, who differ from customers in China. Teenagers in the United Kingdom are likely to use their mobile phones in a totally different way from housekeepers or field representatives. (For examples of differences among customers, usage contexts, and conventions, turn to Chaps. 4, 8, and 9.)

Since the early 1980s, the mobile phone has undergone a transition from a business tool to an upscale accessory and ultimately to a ubiquitous consumer product. Manufacturers who were able to understand and predict customer wants and needs have been the most successful in the game. Their products have followed and established market trends, and they have matched customer needs when it comes to quality, price, industrial design, features, performance, support, brand, and user interface.

Today's mobile phone business is not about selling the uniform black brick to everybody, but just the opposite—designing and delivering the right product for specific kinds of use. It all starts by analyzing and understanding the various segments in each market; some prefer simplicity, we learn, and some are highly cost-conscious. Some want to express their individuality, and some want all the latest technology features regardless of the cost. A key contributor in creating variety and packing personal appeal into products is industrial design. Nokia has actively developed its industrial design practices to be responsive to consumers' lifestyles and the conventions of the fashion industry. To create products whose elements all fit seamlessly together, we also must tailor the user interface to customer categories. Presently we are applying six main customer segments: experiencers, impressors, controllers, maintainers, balancers, and sharers.

Table 1.1 correlates the main consumer segments at Nokia with the corresponding global product categories and their respective user interface attributes.

Nokia's customer segmentation model is a continuously updated view of the global marketplace. It is based on extensive consumer research conducted by both the internal and external marketing research organizations, including trend analysis, sociological background research, and future watch activities. Evolving customer segments drive the evolution of the product categories in the company's product strategy. These product categories are global, but their applicability depends on the nature of the markets; emerging markets behave differently from more mature ones, and Eastern cultures display different consumer preferences from Western ones. In addition, products belonging to the same global product category frequently diverge with respect to their cellular standards, such as WCDMA (wireless code-division multiple access) in Japan, GSM-GPRS (Global System for Mobile Communication–General Packet Radio Service) in Europe, CDMA in the United States—or by their regional feature set, such as a lunar calendar for the Chinese marketplace.

Defining the Mobile Style

The user interface of a mobile phone extends beyond hardware and software—to the sales package, documentation, customer help lines, and even repair centers. Here we will focus on hardware and software and their interaction. The hardware UI is created by industrial and mechanical designers who lay out keypads, control devices and display modules, design the snap-on cover mechanisms and accessory connectors, and wrap this all into an appealing package with the right shapes, materials, and colors. The software UI is created by interaction designers, graphic designers, and software engineers who define the interaction logic and control key combinations, lay out menu structures, design the graphics and sounds, localize the display texts, and implement it all in embedded or downloadable software. This little-understood enterprise can be divided into user interface style and user interface features.

The UI attributes presented in Table 1.1 are the guiding factors in any UI design process for mobiles, whether the designer is working on a new style or on a new feature. An *expression* style mobile like the Nokia 3330 demands a user interface different from that of a *classic* style phone such as the Nokia 6610, because the customers and product drivers are different. Therefore UI style becomes the backbone of a phone's user interface. It is

Table 1.1 Customer, Product, and User Interface Segmentation

CUSTOMER SEGMENT	Balancers	Controllers
STYLE (PRODUCT CATEGORY)	Expression	Classic
SOME CURRENT PHONES	Nokia 3330	Nokia 6610

USER INTERFACE ATTRIBUTES	Intuitiveness Personalization Simplicity	Efficiency Productivity Backward compatibility

Experiencers Impressors	Impressors
Fashion	Premium
Nokia 7210	Nokia 8910
	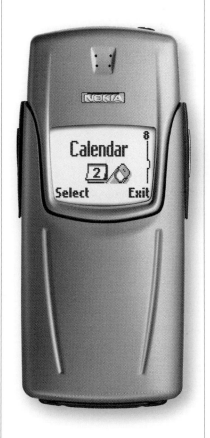
Individuality Appearance Fashionability	Exclusivity Individuality Attention to details

the basic framework for how the user operates the phone, how the menu structure is navigated, and how information is displayed. UI features for the phone—applications such as phonebook, text messaging, or FM radio—come second and are designed to comply with this framework.

This definition of UI style is peculiar to the domain of cellular mobile phones. Donald Norman,[1] the perennial analyst of how things work, discusses a slightly broader domain of interactive and intelligent devices and talks about *smart products* and *information appliances,* respectively. Deborah Hix and H. Rex Hartson[2] approach the concept of user interface style from a broader human-computer interaction perspective and define UI style as follows: "A user interface style is a design framework describing interaction style and objects, including appearance (look) and behavior (feel)."

Nokia's internal definition follows:

The user interface style is a combination of the user interaction conventions, audiovisual-tactile appearance, and user interface hardware.

Let's break that down to its components. *User interaction conventions* describe how input functions are mapped to output functions. *Audiovisual-tactile appearances* define sensory elements of the user's experience, and how those elements are used in accordance with user interaction conventions. *UI hardware* includes display modules, keys and keypads, vibration motors, speakers, sensors, and anything else that makes UI conventions functional and appearances physical. Traditionally, UI conventions are dictated primarily by the hardware control keys and appearance is dictated largely by visual attributes such as graphical layouts. We are now in the process of gradually moving toward a richer user experience with elements such as vibration feedback and polyphonic audio.

We cannot sell UI styles, however, no matter how well informed they may be. The customers are paying for certain features and functionality, and the role of the UI style is to facilitate the creation and integration of a consistent and usable set of these features and functions. New UI styles are legitimately created only in response to some specific customer needs, or to solve real or anticipated problems in an existing UI style. Even then, new styles are not created lightly. An example of compelling customer needs could be a newly identified consumer category or discovery of a specific market with unfamiliar requirements—perhaps the requirement to support a new writing system. An example of anticipated problems could be the incorporation of a new UI technology—say, digital imaging—that has an unintended impact on the original UI.

Clearly defined and documented UI styles are all the more important to a global, multisite R&D community such as Nokia. It's commonplace for a large team of UI designers to be working on a single product and its applications from several continents. These people come from different cultural backgrounds and have different levels of UI design experience. The UI style is the framework for the product's UI that will keep the overall user experience consistent and appealing despite that diversity. Styles are constructed from a specific set of fundamental UI components, and all applications are designed and assembled using these components so that functionally—or structurally—similar kinds of applications look and feel the same.

A Core Set of Scalable Styles

Diverse UI attributes in different product categories imply diverse user interfaces. In many cases this comes down to UI features; for instance, a *classic* phone for business users needs an efficient calendar application, whereas an *expression* phone user probably prefers more a selection of cool games. However, in some cases we need to go deeper than features and applications. To create an appealing *expression* phone for customers who value simplicity, we need to limit the number of control keys on the phone cover. The control keys are an inherent part of a UI style, so the UI style for expression phones should differ from the style used in classic phones.

The focus in this section is on contemporary Nokia UI styles, the two-softkey style called Series 30 (shown in Fig. 1.1), and the one-softkey style named Navi-key or Series 20 (shown in Fig. 1.2). These are the user interfaces some hundred of millions of people are using all over the world every

Figure 1.1 Series 30 style.

day. We are currently applying other UI styles in our product portfolio as well, and our UI designers and usability researchers are continuously working on user interface concepts and designs for the future, but Navi-key and Series 30 form the foundation of our contemporary high-volume product categories.

Series 30

Series 30, the current two-softkey UI, was introduced in the Nokia 6110 and 6190 phones in 1997 and has been applied in evolutionary variants in

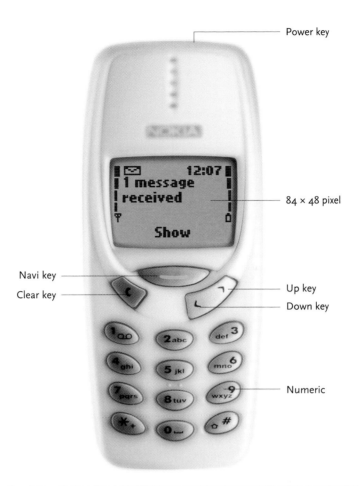

Power key

84 × 48 pixel

Navi key
Clear key

Up key
Down key

Numeric

Figure 1.2 Navi-key style.

a number of phones since then. Series 30 is a descendant of the UI Nokia debuted in the 2110 series phones in 1994. It has two dynamic function keys called *softkeys,* either two or four scrolling keys for navigation, and the green and red call-handling keys (in some markets these are marked with handset symbols and in others with TALK and END labels), power key, volume keys, and a numeric keypad.

Strictly speaking, the power, volume, and number keys are not UI style-specific because they appear in other Nokia UI styles and styles of our competitors. Series 30 has a graphical pixel display capable of flexibly displaying icons, animations, images, and text using different fonts. The original Series 30 UI was designed around a display of 84 × 48 black-and-white pixels. In the Nokia 6210 phone we introduced an evolutionary step, which utilizes a display of 96 × 60 pixels; the 6310s UI evolved further using a display of 96 × 65 pixels; and in the 6610 phone we have Series 40 with a display of 128 × 128 pixels and 4096 colors.

The basic philosophy of the two softkey UI style is to use the left softkey for positive and forward-going operations such as confirming actions, selecting elements, and getting deeper into the menu hierarchy. The right softkey is used for negative and backward-going actions such as closing menus, canceling operations, and erasing inserted text. In addition, the dedicated call-handling keys facilitate intuitive calling from the phonebook and fast and efficient handling of one or more simultaneous phone calls.

Navi-key

The one-softkey UI, Navi-key, was first introduced in the Nokia 3110 phone in 1997 and has been used in a number of Nokia phones since. This UI style has one softkey, a CLEAR key, up and down scrolling keys, a power key, and a numeric keypad.

The philosophy of the Navi-key UI is to be straightforward and intuitive by offering the function most likely to be used in every situation on the softkey. The CLEAR key is used for backstepping and character erasing, and in general for negative and backward-going actions.

To reflect the conscious effort it invests in human-centered product design, Nokia employs usability buzzwords such as "ease of use" and "user-friendliness" in its marketing materials and with product launches. With the Nokia 3110 mobile phone and the Navi-key UI, we raised the visibility of the UI even further by naming and trademarking the Navikey.* An

---

*The 3110 product announcement from Nokia (available at *http://www.nokia. com/news/news_htmls/nmp_970312b.html*) presented the user interface as the most distinctive feature of the new product: "The most distinctive feature of the new Nokia 3110 is its smart menu system. The smart Navi™Key allows fast, one-button access to the functions of the phone, another industry-first innovation from Nokia. Upon pressing the Navi™Key, the phone guides the user through the features, providing fast and easy operation."

easy-to-remember name given to a tangible UI element let us transform the abstract concept of usability from something hard to understand into a concrete product attribute. The message to consumers was as follows:

Anybody can master this phone since it is operated with only one key.

Giving a name to a central usability component focused the whole UI and the product itself on it. Apparent usability was conceptually increased. Trademarking UI elements also makes it easier to employ them as selling points and sales arguments.

There has been some discussion in the global HCI community about establishing "good usability" labels to be awarded to products with proven levels of user-friendliness. In the case of the Navi-key, Nokia combined that approach with a strong marketing message to communicate a concept that reminds customers of ease of use, that they can ask for by name, and whose presence in products is easy for them to verify, even with a very superficial search.

The number of control keys can influence a potential customer's purchase decision. Customer behavior indicates that with the Navi-key user interface style we are optimizing the perceived usability of the phone without unduly compromising "real" long-term usability. Since Navi-key has fewer control keys, some of its features are more tedious to access than comparable features in Series 30, as they need to be invoked through a longer key press sequence. Features and functionality in *expression* style phones are not as rich as in *classic* or *premium* phones in general. Yet market research, user studies, and direct feedback indicate that Navi-key UI users are generally very satisfied with the level of functionality in their phones. We also know that there are a lot of customers who require and respect the flexibility and efficiency provided by two softkeys. True customer segmentation seems to benefit from user interface segmentation.

In theory, we could implement almost any kind of feature or any kind of application, no matter how complex, in almost any UI style. However, UI design, like overall product design, is always a compromise with trade-offs, and in reality the situation is not so simple. At Nokia we often refer to an informal internal concept called the "usability knee" (see Fig. 1.5) to illustrate how each of our UI styles has a breakpoint in the usability-versus-functionality scale. Breakpoint is reached when features get sufficiently complex. We can recognize a rough continuum of usability-critical features and order those on a complexity scale. The criticality of the features is obviously relevant only within the framework given by the UI styles. Sometimes the "knees" are easily discernible while UIs are still on the designer's screen; others are revelations in later-phase usability tests and inspections.

Some critical features are

- *Handling multiple phone calls.* Cellular standards (e.g., CDMA, GSM, WCDMA) all require the mobile terminal to support several simultaneous phone calls. The user must be able to accept or reject incoming calls while already on a call and must have the option of putting active calls on hold, be able to move back and forth between calls, and be able to set up conference calls. Users must have a choice between terminating calls one by one or all at once. Exploitation of these features is culture-, market-, and user-specific; in some calling cultures it is considered rude to leave the other party waiting, while in other cultures it is standard practice. Use of the multiple-call features naturally increases the operators' revenues, as callers consume more airtime. Note that even though Navi-key has highly intuitive core functionality and is praised by its users, it does not support multiple phone calls or other more complex features on the same intuitive level. There are no familiar green and red keys to handle phone calls. All the call-handling functionality is accessed with the one softkey—a key also used for all other functions. To be accurate, a customer with a Navi-key phone can reject an incoming phone call by pressing the CLEAR key, but this is best described as a hidden shortcut as there is no label to indicate the functionality. If the user wants to check a name or number from the unit's phonebook while engaged on a call, the single softkey must provide access to both call-handling and phonebook management functions simultaneously. This kind of complexity leads to long function lists that are slow to use and potentially confusing to navigate.

- *Advanced phonebook.* When the only functionality in the phonebook is to initiate a call, Navi-key is perfect; you can scroll the list with the up and down arrow keys, highlight one, and then simply press the Navikey. However, when designers start incorporating functionality for adding, editing, and removing memory entries; recording secondary phone numbers and email addresses; copying entries between memories; sending electronic business cards; and clustering names into calling groups, then the one-softkey approach starts to become truly cumbersome. If the single softkey must always present a long list of available functions, then the most important function—making the call—is no longer easily accessible. The two softkey UI provides a more flexible and scalable UI platform for advanced applications, as it is equipped with keys for initiating and terminating phone calls and also with keys for navigating the phone's menu structure and performing actions. On the other hand, Series 30 may look a bit intimidating to some customers precisely because of the number of control keys. Series 30 will in turn hit its usability knee in Internet browsing and time management applications that require efficient navigation and access to an even larger number of functions.
- *Time management functionality.* Time management with a small-screen mobile device is about having a basic overview of one's tasks over a given period of time, a useful way to display them, and the ability to manipulate them in a rudimentary manner. The limited 84 × 48-pixel display resolution of the original Series 30 UI style will not permit informative displays of the user's calendar (see Fig. 1.3). However, the 128 × 128-pixel display in Series 40, together with the four-way navi-

Figure 1.3 *Original Series 30 calendar.*

```
┌─────────────────────────────┐
│  Jun 2002═══════════wk 23   │
│  Mo Tu W Th  Fr Sa Su       │
│                     1  2    │
│  ▓3▓ 4  5  6  7  8  9        │
│  10 11 12 13 14 15 16        │
│  17 10 19 20 21 22 23        │
│  24 25 26 27 28 29 30        │
│                             │
│  Options           Back     │
└─────────────────────────────┘
```

Figure 1.4 Series 40 calendar.

gation keys, makes it possible to display a month at a time and conveniently navigate between days (see Fig. 1.4).

- *Internet browsing.* The user experience of browsing an Internet service with a mobile device is relatively remote from the same experience with a personal computer. The display is tiny, familiar point-and-click direct manipulations and navigation are gone, and there is no efficient text input mechanism. Enhanced display resolution in Series 40 UI styles is a minuscule improvement over the original Series 30 display. The Series 60 UI style with its "scroll and click" joystick makes the navigation and selection tasks easier, and more powerful.
- *Rich call functionality.* The 2.5G and 3G (2.5- and third-generation) cellular networks offer more communications bandwidth than do the 2G networks.* This bandwidth will be used to enrich the traditional voice-only service with, for instance, videotelephony, multimedia messaging, and videostreaming. The relatively large color display of the Series 60 UI is requisite for making rich call applications attractive.
- *Text input.* In the early days of mobile telephony, you just keyed in the digits of the phone number. Gradually phones started to store names and numbers in internal memory. Today we are seeing an enormous

---

*2G GSM high-speed circuit-switched data (HSCSD)—up to 57.6 Kbps; 2.5G GPRS—up to 171.2 Kbps; 3G WCDMA—up to 2 Mbps. For further explanation of this technojargon, please see Glossary.

amount of text messaging traversing the wireless networks; in May 2002 there were 24 billion text messages sent globally in GSM networks alone. (See www.gsmworld.com for updated figures.) As phones get "smarter" with email access and Internet browsing capabilities, the need for efficient and convenient text input gets more pressing. The traditional mobile phone keypad can be tweaked to support text entry at about 15 words per minute, but that's far from the 50 words per minute attainable with a decent PC keyboard.* The Communicator UI packs a miniature QWERTY (standard typewriter keyboard layout) keyboard into the product, but the clamshell "minilaptop" form factor may not appeal to all customer segments.

• *Office applications.* The Nokia Communicator is compatible to some degree with Microsoft Office applications. When it comes to advanced content formatting, however, the Communicator falls short. The physical form of a handheld device, where the keyboard and display are significantly more cramped than in PC counterparts, also restricts the usability and utility of portable office applications.

Customer needs are as different as customers. Many phone users neither need nor request highly sophisticated Internet or new media applications. These people can manage well, and perhaps better, with a simpler UI style that supplies a higher level of usability in core features such as voice calling and phonebook applications. Users who ask for more than just voice calling also need high usability in their favorite applications. Thus, usability is not a customer or product segmentation factor; usability needs to be good for all customers in all products. Delivering satisfactory usability for a variety of phone features is possible only through the application of different UI styles in different product categories.

---

*Studies conducted at Nokia Research Center reveal the following text input speeds: conventional mobile phone keypad at 8 to 9 words per minute (wpm); mobile phone keypad equipped with predictive text input at 20 wpm; small personal digital assistant (PDA) device (like Nokia 9210 communicator) keyboard at 35 wpm; standard PC keyboard at 50 wpm.

User Interface Style Evolution

> It is not the strongest of the species that survives, nor the most intelligent, but the most responsive to change.
>
> Charles Darwin*

In the good old days we started UI design work from scratch with each new mobile phone model. Product development took longer back then; there were fewer phone models, models had fewer features, and the product renewal cycle was slower than today. Those conditions allowed us to go out to study what the users wanted, and then come back to the labs to clean the whiteboards, do some designs, create prototypes, and prepare for a world tour to test the prototypes. After that it was iterate and reiterate until an appealing UI eventuated. The revolutionary Nokia 2110 UI was made in this way, as were the Nokia 3110 and 6110 phone user interfaces already mentioned.

Things have changed (to understate the matter) since the early and mid-1990s. In the year 2002, Nokia introduced more than 30 new handset models. The design time for a new product is now counted in months instead of years. In a fast-paced product development project, it is sometimes challenging to maintain a true human-centered approach. What makes it possible is an adherence to planned evolution instead of radical revolution.

In a sense we have invented the de facto cellular mobile telephone user interfaces. Nokia is a market leader with roughly two out of every five mobile phones sold in the world being made by Nokia. The two-softkey UI is being copied by many of our competitors, and our Navi-key phones have been the best-selling Nokia phones for a long time. In a situation like this, revolution really may not be the optimal user interface strategy, anyway.

The *UI style portfolio* is the collection of UI styles we have at our disposal at any given time. It is not static since new UI styles or style variants

---

*This famous quote is widely attributed to Charles R. Darwin (1809–1882), but no specific Darwin reference for the phrase can be found. For further discussion see, e.g., *http://www.csuchico.edu/~curban/Darwin2000.html.*

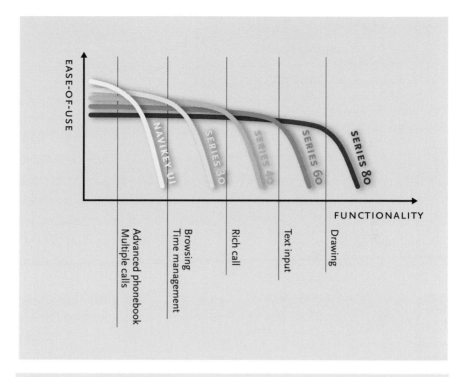

Figure 1.5 The usability knee.

are always being developed. Old ones are gradually fading away or are terminated as they approach their usability knee (see Fig. 1.5). Marketing strategies, technology, or any number of other reasons may also retire a style. Styles are dropped from the portfolio when they are no longer expressive, usable, or otherwise competitive in the changing market-place. On the other hand, we obviously invest more design and development resources in UI styles that have longer-term potential.

Figure 1.6 illustrates the evolutionary development of Nokia's main-stream UI style portfolio (the "big-screen communicators" are not shown in this illustration of the Nokia handset UI portfolio).

For many years, the Nokia 2110 phone with its large display and two dynamic softkeys has defined a baseline for mobile phone user inter-faces. Roughly the same functionality was also available in Nokia 1610 phones, but the user interface was implemented with a smaller display and no softkeys. The Series 30 user interface is an evolutionary step for-

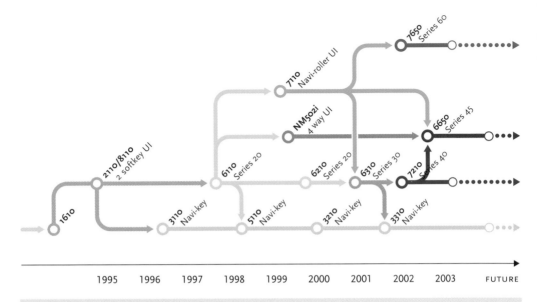

Figure 1.6 UI style evolution.

ward from the Nokia 2110; with Series 30, menu structure navigation was designed in a more intuitive and consistent manner and the two keys for mode switching and backstepping (ABC and CLEAR) became redundant.

The old Nokia 1610 UI style has now been dropped from the UI style portfolio even though it supported the necessary features of its time quite well, and was still used in some analog phones long after the Nokia 2110 UI had been replaced by the Series 30 UI. Gradually we saw that the Nokia 1610 character display was becoming too small and rigid, and the control keys were ill-suited to increasingly complex features. Time passed it by when the fixed-key, two-row display could no longer adapt to the number of oncoming services.

A more recent and perhaps abrupt discontinuity in our evolutionary UI development occurred with Navi-roller. Based on Series 30 with the addition of a larger display and a roller, it was applied in the Nokia 7110, 7160, and 7190 phones. The multifunction *Navi-roller* (see earlier discussion on trademarking UI elements; see also Fig. 1.7) was added to basic Series 30 interaction, so that instead of just scrolling up and down with the navigation device, a user could also select an item by clicking the roller. Navi-

Figure 1.7 Navi-roller UI style.

roller arrived with Nokia's first Wireless Application Protocol (WAP) Internet phones, and the roller device was intended to make Internet page navigation and selection easier. This UI design did not fulfill all of our stringent usability criteria.

Good UI designers have ample amounts of humility in their toolboxes, so we went out to the users and customer care centers, made a thorough analysis of what had gone wrong, and eventually applied what we learned when designing an improved UI style, Series 45, first introduced in the Nokia 6650. Some usability improvements devised as a direct result of the "Navi-roller usability autopsy" project have also been incorporated into the Series 30 UI style displays and layouts.

The reason behind the lack of UI design consistency was that the roller device was added to the Series 30 UI style before we could claim a complete understanding of its implications. The product development team wanted to install the roller because it was seen as a good answer to the "How do I point and click when browsing?" question. We believed that since the roller "just replaces the up and down scroll keys," integration with the Series 30 UI would be no problem, and that proved true in the browser application for which it was proposed. Elsewhere in the UI individual designers started to apply different solutions; in some applications the roller was used to make selections from the menu, and in others it was used to bring up a long list of available functions. This, we learned, was almost like having both the left and right mouse clicks of the Windows graphical user interface (GUI) mapped to the same key. Users struggled to learn the behavior of the roller click. The situation was further deteriorated by the fact that there was no room on the display for a label next to the roller device indicating what would happen if you pressed it. Our internal studies indicate that it takes some time for users to get accustomed to the softkeys with their dynamic behavior, and here we had brashly introduced a highly dynamic key with no instructions!

One happier example of style evolution was the NM502i phone for the Japanese market. It evolved modestly from the Series 30 UI by adding four-way navigation keys and a larger display.[*] Series 60 is the UI style for the Nokia 7650 imaging phone and its successors. Series 60 takes an evolutionary step (or two!) from previous UIs by introducing a larger display, PDA-style multitasking applications, and a more graphical look and feel in the UI. Within the scope of this chapter, where our concern is high-volume product categories and their incremental UI solutions, we can't profitably delve into detail on Series 60. The interested reader is encouraged to visit *www.forum.nokia.com* for more information on the Series 60 platform, which is a complete Symbian[†] smartphone reference design for manufacturers to license and integrate into their own smartphone devices.

---

[*]Laying the foundation for Series 40.
[†]Symbian is a company created jointly by Psion, Nokia, Motorola, Matsushita, and Ericsson to develop and standardize mobile wireless device operating systems.

Benefits of Evolving UI Styles

At the time of writing, Nokia is finalizing the UI style for its first 3G mobile handsets. This style, called Series 45, builds on the proven Navi-key and Series 40 interaction principles, borrows some graphical elements from the Series 60 UI style, and fixes the problems we encountered with our Navi-roller UI. It solves the unlabeled selection key problem by introducing three clearly labeled softkeys.

The familiar two-softkey Series 40 UI carefully morphs into three softkeys (see Fig. 1.8), two of which perform like Series 40 OPTIONS and BACK softkeys while the third, in the spirit of the Navikey, provides labeled direct access to the most needed dynamic functions. All the other control keys are inherited from the Series 40 UI style. This UI is introduced in the Nokia 6650 WCDMA phone and will gradually replace the Series 40 UI style in our most feature-rich mass-volume phones.

User interface style evolution means gradual or incremental improvement in the UI style, as was the case when the original two softkey UI evolved to Series 30 and Series 40. Through controlled evolution, we can leverage an existing UI style and keep it competitive for a long time. Most of our UI design and development work today is evolutionary development.

Sometimes, as we've allowed, a clear customer need is driving a step

Figure 1.8   Series 45 UI style with three labeled softkeys.

in UI style evolution; it is more convenient, for instance, to read and write a full text message on the display instead of scrolling back and forth to see it in several chunks. At other times a technological development we want to promote nudges the UI in a new direction. For instance, improvements in display technologies and greater bandwidth in 2.5G and 3G wireless networks will justify new UI elements such as color displays, and these can then be leveraged for services like multimedia messaging. And sometimes it is a standardization body that is driving UI style evolution.

An evolutionary approach to UI development has practical advantages for the mobile phone manufacturer, too. Small improvements are easier, faster, and cheaper to implement than large ones. It is also easier to turn back, if the chosen solution is not well received on the marketplace.

Consumer benefits of the evolutionary improvement model are obvious. For example, replacing your phone with a new model is easier when the new model has some familiar characteristics (just imagine having to relearn the steering system and gearshift routines for every new car you drive). When we succeed in the evolutionary approach, that is, when we have taken the right steps, customers should see evolution as improvements, and not as disruptive changes. Watch any customer trying out a new phone— comments such as "Oh, they changed the softkey functions" and "This message editor looks funny" may sound a bit negative. We would prefer hearing "Oh, so now I can do this, too!" and "This is what was missing for text messages" instead. As most people already own a mobile phone in the developed markets, it has become extremely important to support replacement customers by providing a smooth evolutionary path from one product generation to the next. It is far easier for the customer to transfer from one UI style to a variant of the same, rather than to a completely new style.

UI style evolution takes several paths. The most straightforward one is to improve display technology, resolution, or color depth. Improvements of this nature will change the visual appearance of a UI style without changing the basic interaction logic. Shrinking the display size or resolution may have desirable effects on product cost and size, but downsizing visible content will displease users accustomed to better. As a result, the downsized product is typically offered to new customer segments instead of customers with the previous model.

Increasing display resolution pays off in two ways: (1) content, both textual and images, can be presented more sharply and clearly, and we thus increase legibility and appeal; and (2) more content can be shown on the screen. The first payoff is the biggest, however. The main usability issue is to reduce scrolling and support the read-ahead function, thereby rendering the features and applications faster and more convenient to use.

The prominence of display in defining small-screen UI style is very different from that with the PC; Windows or Macintosh GUIs scale automatically and perfectly whether you use a resolution of 800 × 600, 1280 × 1024, or 1600 × 1200. In the mobile device arena, however, the UI designer is not so blessed. A 96 × 65-pixel array is basically the smallest display you can use for a calendar view of one month while retaining at least some aspects of good visual design. Any time you have more pixels—say, 128 × 128—then you're all but obligated to use them, which means that you'll have to design new display layouts and probably tweak functionality (now that you can use proper headings, new icons, or scrollbars). For an example of what we mean, please refer to the two calendars depicted in Figs. 1.3 and 1.4.

The keypad can also be evolved. However, each UI style is built on central control keys, and modifying those may result in a completely new UI style instead of a variant. Changing them typically affects the interaction logic of the UI style, meaning that users must learn new skills for their phones, whereas keys dedicated to a more specific use, such as keys for mobile browsing, can be added, removed, or changed without fundamentally changing the underlying UI style.

It's always possible to vary key mechanics, though. This kind of change (e.g., replacing scrolling keys with a roller) doesn't have direct impacts on the interaction logic of the UI style, unless you introduce new functionality as we did with Navi-roller. But ergonomics play an important part in determining user experience, and if the control devices are changed they must also be evaluated—the old usage logic may not be in harmony with the new mechanics. As we defined it in the previous section, a UI style is made up of interaction conventions with specific UI hardware; changing one feature may not be possible without changing the other as well.

Working on creating a new UI style, or in our terms on UI style evolution, is a multidisciplinary task. UI style design requires competencies in interaction design, visual and graphical design, ergonomics, industrial design, end-user and usability research, market research, linguistics, localization, marketing, software implementation, competitor product analysis, and consumer trend analysis, to name just a few. Because of the evolutionary nature of UI style development, it also requires an understanding of both the company's product strategy and the existing user interface portfolio.

Convergence and Standardization?

In the early 1980s there were a number of competing operating systems and user interfaces in the personal computer business. The weakest of them have gradually been phased out of the game, and now the pie is split between two or three major players. Is something similar going to happen in the mobile phone business? What conditions or factors drive an industry to accept de facto standards?

Mobile phone user interfaces are gradually opening up and getting less proprietary through the proliferation of wireless Internet services and applications. Various browsing technologies and Java make it possible for application developers and service providers to reach hundreds of millions of potential customers globally. Contemporary and future smartphones are built on standardized reference platforms like the Microsoft Smartphone 2002 or the Symbian OS, which make it possible to develop native applications for these devices.

The cellular mobile telephone industry is becoming more mature. That fact conspires with the introduction of wireless service and application platforms to promote the evolution of the user interface. Revolutionary development steps may not be so acceptable now that more than half of our business comes from replacement customers who have already used a mobile phone and gotten used to it. These people want their next mobile phone to work roughly like their previous model, and they don't want to spend time learning how to use a product over again.

Application developers and service providers also want a reasonable degree of stability from the wireless device user interface platform, so their services can run on as many handsets and handset generations as

Figure 1.9 Arabic short message service.

possible. Portable runtime environments such as Java facilitate this, while native application development is still needed in performance-critical applications such as games.

Cellular mobile telephone development is becoming more software-focused. Product release cycles are gradually becoming shorter and tenser. Designing an ever-increasing number of handset features in an ever-shrinking development cycle can happen only where the underlying UI platform is reusable and scalable. We've seen that user interface styles can be reused with comparative ease to create variants of a theme. A user interface style developed for a GSM product can be applied in a CDMA, PDC, TDMA, or a WCDMA product, and vice versa, when applicable. A user interface style developed for an expression category phone can be reused in a fashion category phone. UI-style-related modifications are incorporated as needed; creating a phone for the Arabic or Hebrew (Israeli) markets, for instance, requires us to design text presentation and text input around the appropriate reading and writing conventions (see Fig. 1.9).

User Interfaces to Come

The cellular mobile telephone user interface has evolved since the early 1980s without strong de facto or other standards. Where related standards exist, such as the ITU standard* for the alphanumeric keypad layout, they are indeed widely applied. In the absence of standards, user interface conventions have arisen by way of response to the complexity of the user experience.

The latter consists of handset and network features, applications, information visualization conventions, menu navigation and function access, and the "feel" of the device in use. The items in this list relate to human behaviors that are largely *learned.* We continue to believe that the user interface should be developed through evolution, not revolution. We also recognize that manufacturers won't stop introducing revolutionary UI concepts for handheld gadgets, and attractive elements of these concepts may eventually find their way into high-volume mainstream mobile communications devices. The mobile phone user interface will evolve through the push and pull of new UI conventions, technologies, and design improvements brought forward, for diverse and sometimes conflicting reasons, by device manufacturers, application developers, and service providers.

There are already several changes in mobile phone UI development that will influence UI styles. Many of them also crop up in related markets such as PDAs, games, or desktop GUIs. Even though we expect gradual change to prevail over completely novel UI solutions, the limits of evolution and the flexibility of our UI styles will be challenged by some of these trends.

- *Color displays.* Markets like that in Japan have extensively transitioned to color displays. At the time of writing, a similar transition is taking place in Western markets. This development is very reminiscent of growth patterns in the laptop computer business and later in the PDA industry.
- *Input and control device proliferation.* Many phones are now outfitted with scaled-down joysticks. We introduced the roller. Some large-screen smartphones incorporate touchscreens. All these improvements

---

*ITU Recommendation E.161 (02/01), *Arrangement of Digits, Letters and Symbols on Telephones and Other Devices that Can Be Used for Gaining Access to a Telephone Network;* see *http://www.itu.int/itudoc/itu-t/rec/e/e161.html.*

are needed to support more *direct-manipulation* user interfaces. Direct manipulation, in turn, supports content-intensive applications and services such as imaging and multimedia messaging. From a UI style standpoint, it is not at all obvious whether a direct-manipulation UI controlled by a pointing device can be designed based on conventional menu navigation UI styles. Our UI concept creation experiences indicate that totally new UI style types may be needed if and when direct-manipulation and larger screens arrive in mobile devices. Their advent, however, raises an even bigger question: What is the right balance between portability and direct manipulation on a large display? It may turn out that consumers do not want to carry a device large enough to support a true point-and-click direct-manipulation interface.

- *Higher-quality audio.* Both voice control and audio output will improve. Voice control will expand from dialing preprogrammed names in memory to speaker-independent digit and name dialing. When it does, more freeform voice commands will be possible. It is good to keep in mind, though, that voice control has been seen as one of the next-generation UI technologies and post-GUI silver bullets for a very long time. What we mean by "improvement" today is that audio output is in the process of migrating from monophonic beeps to polyphonic tones, Musical Instrument Digital Interface (MIDI), and stereophonic audio.

- *Seamless user experiences.* Bringing the mobile Internet to the masses requires services that can be accessed intuitively and seamlessly via the terminal. Ideally consumers should see no difference or degradation in their Internet experience resulting from the source of content, be it the device, an operator-provided network service, or a freely accessible Internet service.

- *Personalization.* Handset personalization is in full swing with downloadable ringing tones, graphics, games, and user-changeable phone covers. This trend will continue to grow as we start to see user-changeable color themes and downloadable UI skins, screensavers, and other personalization possibilities in the handsets.

- *Replacement customers.* More and more customers are purchasing not their first mobile phone but their third, fourth, or maybe tenth. These people already know what to expect; they are willing to change brand if the previous model was a disappointment, but happy to stick with it if the previous model was a good fit.

- *User interface platformization.* We may not see the strict separation of device hardware and software that typifies the PC business, but proprietary user interfaces are gradually becoming more open and more standardized to facilitate service and application creation. The development and integration of a new mobile phone is, and is likely to remain, a more complex proposition than a new PC model.

- *New-product categories.* The conventional, mainstream mobile phone user interface will progress in organic increments, but in the emerging product categories there is room for revolution. New interoperability technologies like Bluetooth may facilitate the creation of multipart devices having radically different UI paradigms. New customer segments may also have radically different customer needs, and introduce new usage scenarios.

There is no dominant design in cellular mobile telephones user interfaces, but, as a market leader, Nokia has created user interfaces that come closest: Navi-key and Series 30. The maturing marketplace and the needs for more open UI platforms place special constraints on our UI development roadmaps. By strategically balancing the level of evolutionary and revolutionary UI development, we can engineer hundreds of millions of positive user experiences—or else we can fail miserably. Naturally the "right" balance is highly dependent on customer and product segmentation; in many segments customers prefer a clear continuum from their previous phone models, and in some other markets more radical revolution may, and actually should, take place.

Lessons Learned

Like the user interface design itself, user interface portfolio management is an art of logic. What have we learned working on the UI styles at Nokia?

1. *Product segmentation must be a key input to the UI strategy.* Segmentation is a well-known and frequently applied approach in

consumer markets, but creating truly appealing products requires that segmentation reach all corners of product development, including the user interface. Your UI designers must know your users, how and what they think, and what kind of products the company is planning to offer them.

2. *You need a manageable set of user interface styles.* Yes, perhaps you should design one UI style for Brazilian schoolgirls and a completely different one for Finnish lumberjacks, but eventually you just won't have the design and implementation resources to focus on everybody. Too many UI styles mean that you lose the advantage of streamlining rapid product variant creation and consistency over different product generations. Too few UI styles mean that a significant part of your customer base won't be happy. Nokia's current approach has been to cover most of the high-volume product categories with two UI styles and their variants—Navi-key and Series 30—and then introduce new UI styles in emerging product categories to probe the markets. This has proved to work since the mid-1990s.

3. *User interface styles must have a sound foundation and still be flexible.* You don't know now what features you'll have to integrate into the next-generation product, so the UI platform must be solid and scalable. Longevity of a UI style also makes it easier for the replacement customer to stick with the familiar brand name. On the other hand, UI styles inevitably have to accommodate new features, work with new technologies, and allow personalization. Nokia's approach to UI styles mandates dynamic softkeys and scalable screen resolutions, and it has therefore proved to be very flexible with respect to new features and displays. The underlying software architecture is as solid as we can get it, to support about 50 different display languages today.

4. *Watch for the revolutionary opportunities.* The marketplace is continuously changing as operators merge or vanish, and as new customer generations and segments appear. User interface technologies are developing, making completely new product categories and form factors possible as they do. You'll have to be continuously alert to predict how existing UI styles will cope with

this (watch for approaching usability knees) and to kick off new development work in time to be among the first to reap the benefits of new opportunities.

What will the future be like? Nokia will definitely continue to evolve and improve the Navi-key, Series 40, Series 45, and Series 60 user interfaces for mainstream product segments while introducing new and sometimes more revolutionary user interfaces, just as we did with the Series 80 Communicator UI. Beyond our walls, the marketplace grows and changes, new customer segments are born, and old UI technologies give way to new ones. A solid UI style backbone is essential for building good user interfaces for the mobile information society.

References

1. D. A. Norman, The Invisible Computer: Why Good Products Can Fail, the Personal Computer Is So Complex, and Information Appliances Are the Solution. Cambridge, MA: MIT Press, 1998.
2. D. Hix and H. R. Hartson, Developing User Interfaces: Ensuring Usability through Product and Process. New York: Wiley, 1993.

Applications

Telephone

Contacts

Logs

Messages

Camera

Images

Calendar

Services

Profiles

Options

Exit

Seppo Helle, Johanna Järnström, and Topi Koskinen

CHAPTER 2

Takeout Menu
The Elements of a Nokia Mobile User Interface

In this chapter we discuss the elements of Nokia mobile phone user interface: the menu, softkeys, shortcuts, and localization. How and why were these solutions adopted, how are they applied in our designs, what are their benefits to the user, and what kind of problems may they involve? We present detailed examples of the design implications set by the small product size and the wide consumer range.

The Ultimate in Mobile Computing

Meet the Apollo Guidance Computer, a rather special piece of equipment designed and built for use in Apollo spacecrafts during the moon flights of the late 1960s and early 1970s. Its processing power was very modest compared to today's equipment, but it remains a recordholder in its own arena; no other computer has ever been used by humans so far from home and traveling so fast.

The programs in the Apollo Guidance Computer (Fig. 2.1) performed star navigation measurements, controlled the spacecraft's orientation and propulsion systems, guided lunar descent and ascent maneuvers, coordinated the rendezvous of the spacecrafts, and performed numerous other tasks. At that time, using computing equipment was not an everyday job. Only the most important tasks were worth the formidable price of the

Figure 2.1 User interface module of the Apollo Guidance Computer. The right-hand display panel contains seven-segment numeric displays; the left-hand panel is for warning lights. The computer itself was housed in another, larger unit. Two of these were launched into space for each lunar mission: one in the command module, another in the lunar module.

hardware. One needed a fair amount of training to be able to utilize a computer; and often the user was also the programmer.

The user interface of the Apollo Guidance Computer was built into a display-keyboard module (DSKY, or "diskey" in astronaut jargon). It contained 19 push buttons, three 5-digit numeric data displays (consisting of seven-segment digits), three 2-digit code displays, and some 15 warning and indicator lights. The number of buttons and the display's data capacity in today's mobile phone are surprisingly close to those of the Apollo computer—and indeed, in both cases the need to save space and weight is a critical design requirement. [The giant leaps in technology since the early 1970s become clear when you know that a 31.7-kg (70-lb) processing unit and a 7.9-kg (17.5-lb) user interface module were actually considered a lightweight design back then.] There was another similarity in respect to modern mobile phones—in the Apollo case, too, the users were strictly users, not programmers.

Like us, the Apollo astronauts had buttons to push and a screen to watch, but their tasks were very different from ours, and that difference is clearly reflected in the designs. Because astronauts receive extensive training, intuitivity and learnability are not as important for them as efficiency is—quite the opposite of a consumer device aimed at millions of untrained individuals. The ideal UI design could not be the same for two such different types of equipment even if the processing technologies inside the boxes were.

In the 1960s, integrated circuit technology was still taking its first steps, and it limited the amount of memory and processing power available for computing. Within the memory space of 2 kilowords of RAM and 32 kilowords of ROM built into the computer, there were few other alternatives than typing the commands in as numbers. The Apollo computer used a peculiar command interface where two-digit command codes were entered either as VERBS or NOUNS. The VERB key told the computer to interpret the numbers as an action command—for example, VERB 49 meant the initiation of a crew-defined maneuver. The NOUN key cued the computer to interpret the numbers as a parameter, or as the object to which the action applied. One VERB code was used for loading PROGRAMS; there were about 70 of those in use. Astronauts memorized hundreds of numeric codes that helped them know what was being done, and their flightplans were also equipped with checklists to help them keep everything in order. This system worked quite well for the task, and it was optimized for weight and space requirements.

Takeout Menu

The vanilla home stereo has separate knobs for volume, balance, bass, and treble, and push buttons for selecting the radio, CD player, or other sound sources. The kitchen stove has one dial for each burner. This one-function/one-button principle was used in early mobile phones as well, but expanding feature lists and shrinking electronics have driven phone UI design toward other solutions. Command interfaces would get the job done, but they are better suited for moon walkers than pedestrians.

The accepted solution turned out to be the menu. It found its way onto stationary computers first, and then also proved useful for small gadgets carried around far and wide such as the early mobile phones (see Fig. 2.2). Today, with the menu as a cornerstone of the Nokia phone UI, it is fair to define that interface as a *menu UI*.

The fundamental idea of the menu is to present functions in a structure so well organized that even a first-time user can find and operate them all. A menu UI reduces the learning curve considerably by comparison to a *command UI*. Users don't need any advance knowledge of available functions in each application and how to invoke them. They merely have to learn how to browse the menu and select items from it.*

In Nokia phones, a menu concept consists of the following items, each having its specific purpose, requirements, and representation:

- Idle screen
- Application menu
- Hierarchical submenu
- Option list

Let's take each component in turn.

The *idle screen* is the launch pad for all phone functions. A phone goes to "idle" when it is powered up, and there it returns when calls and other tasks are completed.. The idle screen displays information about the current status of the phone (see status icons illustrated in Fig. 2.3); the quality of the cellular signal, the identity of the currently accessed network, the amount of power left in the battery, the profile name, and so on. It can also display notifications about received messages, missed calls, and other events that the user may not have noticed. Finally, the idle screen

---

*It should be noted that we are describing the classical mobile phone user interfaces, and not devices categorized as personal digital assistants. Still, many of the issues raised are relevant for a broader range of products than just phones.

Figure 2.2 One of the first hand portable mobile phones, the Mobira Cityman 900.

Figure 2.3 A status icon in a Navi-key or Series 30 style is just 6 pixels high; the width can vary. This few pixels means efficient screen usage, but it also calls for minimalist graphic design. One can almost say that when drawing such a small icon, it's possible to try all possibilities, although 2^{49} is really too large a number to do it literally. Some metaphors just don't work in such a tiny space.

grants access to the applications menu—one key is always reserved for opening the menu from idle.

Designers know not to pack an idle screen with too much information. As the most frequently seen view of the phone, it should be calm while distinctive. As the main view to the phone's interactive functions, it should also act as a kind of home page—a place for the user to personalize the device.

The *applications menu* (see Fig. 2.4), also known as the *main menu*, lists the main functions and applications of the phone. There are the traditional phone applications like phonebook and call registers, and others that vary with the phone model such as the calendar, ringing-tone composer, and games. The applications in the main menu will change from product to product, but the essential applications are kept in their "traditional" places.

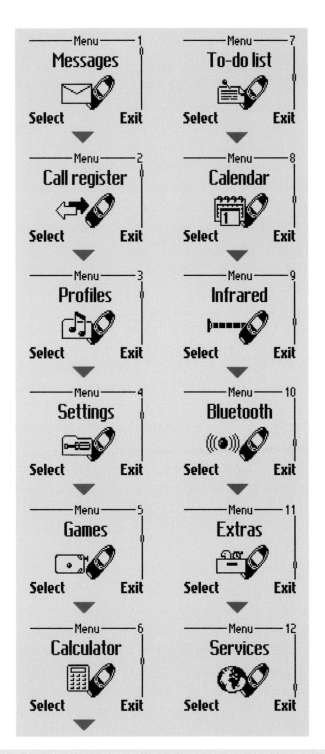

Figure 2.4 Applications menu in Series 30 style.

The purpose of an applications menu is to give the user an overall idea about what the product can do. We think it's essential that the user can learn the product's possibilities soon after switching it on. The main menu should have a reasonably limited number of items; much more than 10 are considered excessive. The items should have communicative names presented in a straightforward fashion, one by one, to increase the user's confidence. We chose to combine text and graphics in the main-menu layout; text communicates the application's name in an unambiguous way while the graphic can emphasize some characteristics of the application and give extra visual appeal to the menu.

Applications are often constructed using *hierarchical submenus* whose appearance depends on the kind of items it contains:

- Higher-level menus, and ones with items that contain long texts, use a *full-window layout,* where each item in turn occupies the complete display area (some styles that apply larger screen sizes, e.g., Series 45 and Series 60, use layouts containing a few items each with more than line of text).
- Lower-level submenus are presented in *one-line layout,* assuming that their texts are short enough to fit on one line, so that more than one item can be seen simultaneously. Also, lists that may grow long, like the list of names in the phonebook, are typically presented as one-line items.
- A *setting items layout* consists of a title field and a value field, displaying the current value of a setting, for example, "Ringing tone: Attraction," which tells the the name of the ringing tone that is in use.

An *option list* is used in certain states of the application as a means of accessing various generic and context-dependent functions. It can be seen as analogous to the drop-down menu in computer interfaces. The option list is accessed through a softkey labeled OPTIONS and is not actually a direct descendant in the submenu structure. Options do not take the user deeper into the menu hierarchy, but present a list of alternative actions applicable to the context. In styles such as Series 60, where select function and options occupy separate keys, the option list is not a part of the hierarchical menu tree but a tool to access special functions—usually secondary functions—leaving the menu tree streamlined for the primary function. This arrangement can improve efficiency. In Series 60 the option list can also be given a distinctive look, partly because of the larger screen of the style. (See Fig. 2.5.)

Used judiciously, differences in layouts can deliver helpful information; when the layout is distinctive enough, the user can determine the status of the phone with just a glance at the screen. Whereas most other texts are left-justified, the idle screen and main menu are centered, making them unique and easy to recognize. You can tell from a distance, without actually reading the text on the screen, when your phone is just sitting there idle. If it's not, you might have a closer look and check whether you have received any messages or calls.

Although mobile phone menus look different (see the four styles compared in Fig. 2.6), varying with the amount of information to be presented and its nature, all menus work the same way. Users scroll up and down the list and select an item. There is a way to back out of the list without making any selection and return to the previous state. For these actions the user needs a practical minimum of four keys:

- A SELECTION key for opening a new submenu, opening a menu item, or applying a function to a selected menu item

Figure 2.5 Series 60 option list

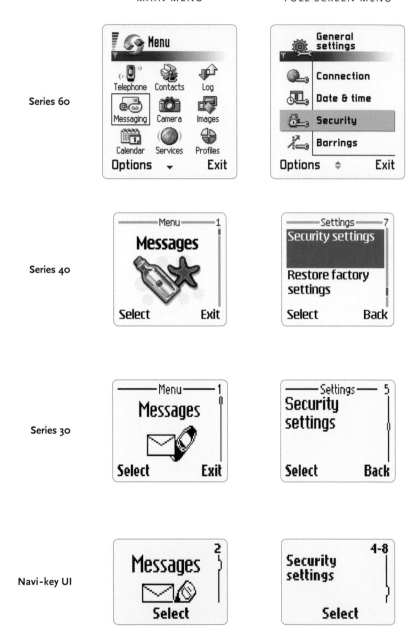

MAIN MENU   FULL-SCREEN MENU

Series 60

Series 40

Series 30

Navi-key UI

Figure 2.6 Examples of mobile phone menu layouts. Our approach aims at giving the hierarchical levels distinctive appearances to help users navigate the menu. This can usually be done, although the available means are limited. In some cases, the designer's options are restricted by the content; if the menu

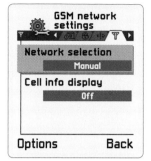

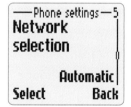

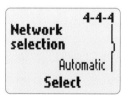

item texts can't be truncated to a couple of short words, the designer may need to choose a menu style that is not "correct" for the hierarchical level. Usually the design should prefer full-screen items in the top levels and switch to layouts with smaller items in deeper levels, but well-argumented exceptions to this do exist.

- A SCROLL FORWARD key for pointing at the next item in a menu
- A SCROLL BACK key for pointing at the previous item in a menu
- A BACKSTEP key for returning to a higher menu level

Depending on your UI style, the SELECTION key or SELECTION key and BACK-STEP key are realized with a softkey. The softkey is a button whose current function reads out on the bottom of the screen (see Fig. 2.7). All modern Nokia phones have at least one softkey. (Chapter 3 guides you through the history of the one-softkey UI style.)

The advantage of a softkey is that it can be used for many different functions depending on context, improving flexibility in the UI without increasing the number of physical keys. As the user can always see the function assigned to the key indicated by a clear written word, usability should be as good or better than permanently labeled keys with their small and often ambiguous icons.

Another way to reduce the number of control keys in a menu UI is to make use of long key presses or double-key presses, but that wouldn't be highly usable. Controls that rely on multiple keys or prolonged user touch must not be used for the most vital operations of the UI. Users would have difficulties in both discovering and adapting to them. Comfortable one-hand use has always been regarded as an important requirement for a product to be used in mobile situations.

Even though the menu-based user interface makes phones simple to access and accommodates new features easily, it has significant limits. Some of those are specific to small mobile devices:

Figure 2.7 The label of a softkey identifies the current function of the key very clearly.

1. The length and depth of menus increase with the number of features (see Chap. 12). The menu structure may become increasingly scattered and clumsy, while access to more peripheral functions, which still may be frequently employed by some users, becomes slower.

2. Menu browsing requires users to scroll through items one by one, looking for what they want. Frequently used sequences can be memorized, of course, and if the menu is small enough, checking items one by one is completely acceptable. It becomes less so with long lists that may not be organized to facilitate the user's search, however. And making the user scroll through a menu while someone hangs on the line for a response is too much to ask. Patience will end and rules will change.

3. On a small screen displaying one item per line, the length of menu item names will influence the visual style of the menu, and because the word lengths depend on the language spoken, menu design and localization are intertwined.

4. As simple as the idea of softkey is, it requires the user to relate up to three UI items with each other: the selected menu item, the softkey label, and the key itself. This doesn't always come off without a hitch.

Sturdy Boosters Wanted

The menu is a good tool for occasional use, but it may become boring to navigate over time. Besides, menus and option lists are getting longer, or the hierarchical structure of applications is getting deeper, or both. Some frustration caused by long and deep menus can be reduced by arranging the menus carefully; important and frequently used features should be accessible from the top screens and at the beginning of lists. Sooner or later, though, different users need different features, and they can't all be first. We have to provide shortcuts for customers who want to speed up their interaction with the phone. A shortcut is an alternative, quicker way to select a function. Shortcuts (see examples in Fig. 2.8) are hidden beneath the visible level of the user interface, which means that there are no labels

IN IDLE

INPUT   **5**jkl

MEANING   Enter '5'

IN IDLE

INPUT   **5**jkl , long press

MEANING   Initiate speed dial to defined contact

IN PHONEBOOK

INPUT   **5**jkl

MEANING   Go to first entry beginning with 'J'

IN TEXT ENTRY

INPUT   **5**jkl

MEANING   Enter 'J'

IN PROFILES MENU

INPUT   **5**jkl

MEANING   Switch to 'loud' profile

IN IDLE

INPUT   **menu** + **5**jkl

MEANING   Go to menu item 'sound settings'

IN TEXT ENTRY

INPUT   **5**jkl , long press

MEANING   Enter number 5 (instead of letters)

Figure 2.8   Shortcuts associated with key 5. Nokia phones are bubbling with shortcuts, but it depends on which one(s) the user applies. The shortcuts provided depend on the current status. In the idle screen, in phonebook scrolling, and in text entry state the shortcuts change accordingly to allow efficient operation for those who take the trouble to learn to use them. The figure lists the possible shortcuts that are related or can be associated with key 5.

on the screens or prints on the keys to guide the user. Shortcuts require learning on the user's part. Therefore, they cannot replace a menu, and we cannot build UI performance measurements on shortcuts alone. A shortcut is an amenity to enhance the experience of the keen and demanding.

A good shortcut does not interfere with ordinary usage, is somehow memorable, lets users feel they are in control, and above all, actually reduces the effort to accomplish something.

An example of a shortcut combining many of these benefits is the redial shortcut, where we use the green SEND key as a shortcut to redial list. When the phone is in idle state, pressing this key once brings up the last dialed number, and a second press sets up a call to this number.* The menu method of performing this task is as follows:

1. Open the menu.
2. Scroll to the call register application.
3. Select this application.
4. Scroll to the dialed calls list.
5. Open the list.

The menu method typically requires about half a dozen presses on various keys, where the shortcut method assigned to the green key requires only two presses of one key. Think of the redial shortcut as a kind of history that the SEND key has collected about its usage. Pressing the key reveals information about how it was used the last time—and in fact a longer history of previous calls can be retrieved by scrolling the list. Thus, the key's main function and the shortcut are related in an intuitively meaningful manner, which makes the shortcut understandable and easy to recall. Another essential shortcut is to give the user an access to the alphabetical list of saved names and numbers directly from idle screen by pressing scroll keys. Such lists should jump to the desired spot in the list when a user enters the name's first letter on the number pad.

To let the adventurous feel the excitement of space navigation without the risk of really crashing into the moon, there is also a general numeric

---

*It should be noted that there are variations in the shortcuts for certain products of certain cellular systems or market areas.

shortcut system to any applications submenu.. It works by pressing the MENU softkey and immediately after that the number of an application in the applications menu, the number of the application's first submenu, and so on. For example, keying the sequence MENU–4–1 in a Series 30-style phone accesses the alarm clock. People are not supposed to know the numeric sequences to all features, but it is easy to memorize the sequences of the few features one uses daily.

When the Rules Change

As Neil Armstrong and Buzz Aldrin approached the moon's surface for a landing, their computer suddenly produced a master alarm, flashing a warning light and displaying the error code 1202. The code was not familiar to the astronauts, so the experts at mission control in Houston had to look at their handbooks and find out what it meant. It was an overload error—caused by the landing radar producing more data than the computer could digest—but it was not fatal, and the mission was still a go. The same kind of alarms went off again a few more times during descent, but as we know, they made it safely down.

While working on the UI design of the first Series 30-style phones in 1997, we realized that our call handling was not as competitive as other manufacturers'. We looked into the reasons, tossed around ideas for improvement, and implemented two major changes, one of which was the call menu. The UI for call handling went through a remarkable change. The new call menu brought call-handling features available into the softkey label on screen during a call.

During a call, the call menu in Series 30-style phones is accessed by the OPTIONS softkey. Previous phones also had a call menu, but it was situated in the middle of the main menu. True, it was logically where all other menu items were, but the situation in which users have to access call-handling features is fundamentally different from a situation where users can take their time and browse. The usability tests of call handling were interesting at this point. When we gave the users the task of setting up a conference call, or making a second call and returning to the original call, the test users initially regarded these tasks as impossible to accomplish. Afterward they were surprised at being able to carry out the tasks successfully with the help of the call menu. We were relieved to have made it down safely.

Not all Nokia phones have dedicated volume keys. In those that don't, the speaker volume is adjusted by manipulating the up and down arrows (i.e., the scrolling keys) during a call.[*] It is easy to do once you know about it, but many users have problems finding this feature in their phones. They try to find it but can't—and this applies even to some dealers! The feature can't be tried without a call; when the phone is idle, the scroll keys open the list of names in the phonebook instead of volume control. During the course of a call people do not explore new possibilities with the phone—they speak.

Menu solutions, like most other UI solutions, aim at helping users learn new operations and find new functions by exploring the product with what they've learned.. The core user interface does not require users to learn anything by heart. These working assumptions follow the well-known UI design principle: *Rely on recognition, not recall.* Everything is presented to the user, who need only choose.

During a call, however, the rules change. Users do not feel relaxed enough to study the user interface. Conversation occupies their attention. Perhaps shifting one's attention from the interlocutor to the device feels impolite; perhaps the fear of failing to best the UI while someone is on line to witness it makes users uneasy. Or perhaps simultaneous human–human and human–machine interaction simply leads to a cognitive overload. Whatever the reason is, interactions required during a call have to be right on the top of the user interface. They must be realized with dedicated keys, or with a menu directly accessible from the idle screen.[†]

We Say It Your Way

Customers today expect to use their phones in their own languages. That expectation is uncommon in most other forms of information technology,

---

[*]Or in Series 60 using the left and right arrows during a call.
[†]The problems of operating the phone during calls as described here are related to basic phones. Smartphones and communicator products are designed specifically to enable fluent operation of different applications in parallel with calls.

and it may well derive from the fact that some major players in the mobile phone business emerged from small-language countries. It also stems from the design decision to make mobile phones easy-to-access consumer products rather than specialized pieces of high technology. Today, it's no longer possible to conquer the world without localization.

Good localization takes many issues into account, some of which are contradictory—everything displayed must fit on screen, terms should be

Figure 2.9 There can be radical differences in the appearance and length of texts that are translated into many languages.

familiar to users, terms should be used consistently across the software and software manuals, abbreviations should be avoided, and the grammatical rules of each language should be scrupulously obeyed.

English is one of the shortest languages, and accordingly most software design is done in English—the corporate language. Translating texts into languages like German or Finnish, though, may result in significantly longer words. Sometimes a language also requires a different construct

using more words than the comparable English expression. Squeezing all essential information into a translation can be quite a battle, complicated by the fact that layout style will be a function of text length. (See Fig. 2.9 for a comparison of menu items in nine different languages.)

The driving principle in localization is respect for the language of the users. Expression is permitted to take the space it needs (within limits), and the UI structure has to be flexible enough to tolerate this. One-line menu items may be preferred in general, but if one line does not provide enough space for full expression of content, even if that's true only in some of the languages supported, designers will accede multiple lines. Acronyms are not accepted, and technical terms are used only if there are no corresponding expressions in common parlance.

The software platform for any mobile phone must support the characters and diacritical marks used in all target languages. Some writing systems have character sets completely different from the Latin character set, and a few languages require the ability to draw text from right to left. An *internationalized* software environment is needed, an environment that allows applications to be translated into all intended languages. Creating one can be a challenging task, though. Many existing platforms have only limited support for languages with requirements very different from those in English.

That said, software support may not be enough. Languages such as Chinese have input requirements so specific that they can significantly influence product concepts. This sort of disparity may help explain why, for example, some parts of the world like pen-based UIs while keypad-based interactions are solidly preferred elsewhere.

Localization was a nonissue in Apollo's case. Everybody involved spoke English, anyway. Well, actually, human language had no relevance to the design of diskey as it defined a language of its own, a completely artificial language optimized for its own comprehension. Today localization is a major part of the work of developing consumer products. In the Nokia classic category phones, for example, the number of different display texts approaches 2000 for each language. To translate them all into tens of languages, following localization guidelines and ensuring that all translations are ready when the product ships, is the larger part of UI implementation work in product programs. (For a case study of localization during UI style creation, turn to Chap. 8.) Figure 2.10 shows the countries with currently localized languages.

How Soft Can a Key Be?

Softkey is not just a key, but a construct with several user interface elements:

1. The key which launches the action when pressed
2. The variable softkey label, which identifies the action in a context-dependent manner
3. The subject of the current softkey action, which is typically an active selected menu item*

For the softkey concept to be understandable, users need to figure out the meanings and the relationships between the elements. Easy as it might sound, the softkey concept is not without a few inherent usability problems.

- *Associating the key with the label.* When we first introduced the concept of softkey, novice users had trouble matching the physical softkey with the label displayed. For some users these problems weren't especially serious, but the observation is unquestionable nonetheless. Typically users are able to overcome this problem and adapt to the softkey UI with practice. It probably arises in the first place because certain users perceive screens and keypads as different domains of the user interface, and find it hard to comprehend UI elements that are logically one thing, but exist in two domains. We have tried with various visual

*The softkey concept does not require the existence of a corresponding menu item. For instance, the idle screen has softkeys but no corresponding elements of this kind. However, softkeys in Nokia menus are typically applied by linking softkey labels and menu items, and the link explains many of the usability problems related to softkeys.

elements to link the screen and keypad parts of the softkey together, but close proximity of the two has turned out to be the most effective way of doing it.

- *Ignoring softkey texts.* While our colleagues at NASA were able to rely on the moon explorers to remember dozens of artificial number codes,

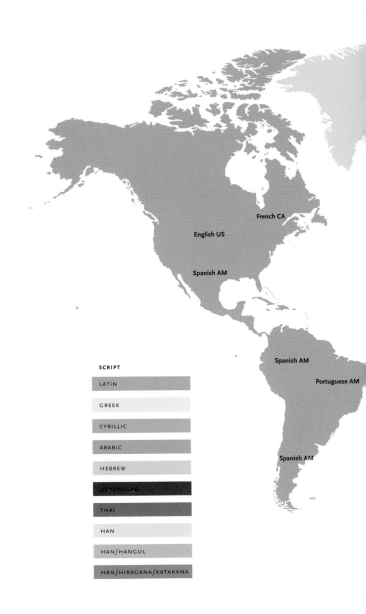

French CA

English US

Spanish AM

Spanish AM

Portuguese AM

Spanish AM

SCRIPT

LATIN

GREEK

CYRILLIC

ARABIC

HEBREW

DEVANAGARI

THAI

HAN

HAN/HANGUL

HAN/HIRAGANA/KATAKANA

we cannot count on today's highfliers to read both of the two words presented to them in their mother tongues on the screens of their devices. When reading menu texts, users tend to ignore softkey texts. When browsing lists, users pay attention to the items on a list but overlook the softkey labels. It's quite common for users to press a softkey

Figure 2.10 World map illustrating all areas with currently localized languages.

without actually reading the label at all, which can lead to errors. What's the upshot of this behavior? On one hand, we strongly require that the meaning of the softkey in each situation approximate the user's expectations. That requirement is best met with a *consistent* design, rather than an extremely context-sensitive one. On the other hand, keeping the softkey static disregards the dynamic and context-dependent potential of the concept. The power of the concept is lost if we can't associate the softkey with the most likely inputs in each situation. Softness has its limits. Only careful design and testing can lead to the best softkey solutions.

- *The priority of the primary function.* The one-softkey style, Navi-key, with its four control keys appears to be a good solution for many simple tasks, but it calls for simple designs overall. When designers can assume that the user wants to perform just one type of action in a given situation, that action can be assigned to the only softkey, and the user should have an easy time of accomplishing the task. But whenever more than one action is available to the user, the softkey has to be labeled as OPTIONS. Any number of actions can then be collected into the options list, but now none of them is accessible by a single key press, which slows down the interaction. It also partly ruins the rationale of the concept. This tradeoff shows in the design of the phonebook. In a one-softkey UI, there is no dedicated SEND key, and for efficiency the softkey is assigned as "call" when the names list is being browsed. Accordingly, number editing or other actions are not available at all in this state. Such a function has to be selected before starting to browse the name list. In other words, instead of choosing a name and deciding what function to apply, the user chooses an operation and finds someone to do it to. The principle of maintaining consistency in the object-action sequence throughout the user interface has been broken here to allow more intuitive access to the most important function.

- *Screen area challenge.* When the length of menu item text becomes a problem, it will be even more of a problem for softkey labels. One short word has to be found in all languages for all softkey texts. Icons would save us space, but the requirement for easy interpretation has been shown to favor text labels.

Giant Leaps

Nokia's mobile user interface is based on a hierarchical menu that is operated with softkeys. We have introduced our set of menu types and the
control keys applied in our basic phones. Menus and softkeys are the logical backbone of the user interface. We have also discussed other UI levels that are related to the menu structure. Shortcuts provide a layer of
interaction that aims at higher efficiency than a menu UI. Interaction during calls also calls for solutions that differ from the hierarchical menu in
being more overt and direct. The logical UI level defined by menu structure must be complemented by a layer of language that communicates
the meaning of menu.

The achievements of the Apollo Guidance Computer were beyond
respectable. Moon flights were one of the most outstanding milestones of
technical development in the twentieth century. Their consequences
were felt in several areas of human life, from technology to global politics,
and the diskey (DSKY) UI was a small but vital part of the technology
behind the giant leap.

What are the parallels between diskey and mobile menu on the level of
social or societal consequences? Are there any at all? With something like
10,500 key presses on the Apollo's computer system, three astronauts
were able to fly to the moon, land there, and come back home—a tremendous achievement in itself. The same number of key presses today could
mean, for example, 35 text messages (160 characters per message, two
key presses per character on average). A present-day teenager could use
those key presses to reach out to another one living somewhere else in the
universe—or maybe next door—and fall in love. That may be one small
step for mankind, but it's a giant leap for the individuals involved.

CHAPTER 3

The Navi-key Story

Origin of the Navi-key Concept

In 1995 the mobile phone business was at one of its turning points. Devices that were previously expensive and exclusive had begun to spread to ever-larger consumer populations, and a new user interface was needed to match the needs of the new customers. This user interface was supposed to be very approachable for first-time users who were neither familiar with nor interested in technology. At the same time it had to provide access to the full scope of functions required by GSM* specification, already considerable at that time.[†]

During the spring of 1995 Nokia also kicked off a project reconceiving user interface for a new business phone user interface of what later would become the Series 30 (two-softkey style, see Chap. 1) first introduced in Nokia 6110-GSM phone. At the time the very secret Communicator project was under way, attempting to combine a personal digital assistant

---

*Global System for Mobile Communications was originally a European digital system for mobile communications. It was first introduced in 1991. Now GSM has become the de facto standard in many regions around the world, serving more than 100 nations. The notable exception is the United States, where adoption of GSM is still in its infancy, and analog networks still dominate. More than 239 million people around the world use GSM networks. Technologically, GSM uses what is known as narrowband time division multiple access (TDMA), which allows eight simultaneous calls on the same frequency. GSM works primarily in three frequencies: GSM 900, GSM 1900, and GSM 1800. GSM 900 system is the most extensively used worldwide. GSM 1900 is primarily used in urban areas in the United States. GSM 1800 is primarily used in urban areas in Europe.
[†]A user interface design project and a team managed by the author, product manager Christian Lindholm, were set up.

(PDA) and a mobile phone. Everybody was very eager to work with these high-end products. A basic phone seemed so trivial somehow. Fortunately the project team had a couple of patient mentors, marketing manager Erik Anderson and design manager Mikko Palatsi, who knew that it would be the basic phone that would make the mobile revolution.

Nokia had just launched an NMT* phone called *Ringo.* Ringo had only a green SEND key for making calls or answering them, and a red END key for terminating calls. Furthermore, it was stripped of most other features, even down to the usual alpha printings on the number keypad, and memory was limited to 10 speed dials. Ringo was the ultimate in simplicity in our industry. It was so simple that people began referring to it as "the bimbo phone," an allegedly humorous gibe at the ignorance of users. We were quite disappointed because Ringo was a very earnest attempt to make an easy-to-use phone for users put off by technology.

Meanwhile work continued on the business phone user interface (UI). Development was based on the two-softkey design of our previous GSM model, Nokia 2110. As we wanted to create two clearly different segments UI-wise, that solution was out of the question. And since the Ringo experiment led to too many limitations on user control, we knew that we had to look for something else for the new consumer phone.

We started brainstorming alternative ideas. *Wilma* was a concept based on a single softkey and a rotating mode key, similar to the ones found on cameras. Another minimalist concept with only three control keys—two softkeys and a clear key—was called *David.* David seemed wonderful at first . . . particularly if you were a power user on the development team. The problem with David was that the meaning of the softkey changed by timeout. We were constantly arguing about the length of timeout before key functions would swap.

Simulations of Wilma and David were completed for a world tour, and once it started, the inevitable occurred. Design manager Mikko Palatsi, the inventor of David, yelled in the usability test backroom behind a one-way mirror: "This is a catastrophe! We'll never make this, they don't get it!" And we didn't. We didn't build Wilma, ether. But although both concepts were buried, there was something that we really liked in the con-

---

*Nordic Mobile Telephone is an analog system used in the Nordic countries.

cept variants. The single softkey of Wilma seemed so elegant. David, in spite of its failure to beat Goliath this time, introduced the idea of a mobile phone UI without dedicated keys for starting and ending a call.

After the experiments and lessons learned from the first round of concept creation, we were able to formulate the primary design drivers that we would try to adhere to.

- As few keys as possible
- No dedicated call handling keys
- As many universally recognized keys as possible

Only a few keys were desirable because that would create an instant impression of ease-of-use at the point of purchase, increase perceived usability, and statistically reduce the number of possible wrong key presses. Minimizing the number of keys appeared simple, calm, and inviting.

Above all, we wanted to get rid of the dedicated call handling keys: green SEND key and red END key. These keys caused confusion for mobile phone novices who followed the logic familiar from landline phones. They tended to press either the red or the green key as if they were power toggles, in order to get a dialing tone. Since that did not work in the GSM network, they became flustered and were thwarted by the fundamental task of placing a call. This was a ridiculous flaw in mobiles. We aimed at getting around the dialing tone problem by removing the receiver symbols, as the green and red handset keys are powerful finger magnets. Users had to learn to punch in the number before trying to contact the network.

Armed with design drivers, our design team started playing around with several key combinations using a set of frequently performed tasks as a testbed—tasks such as saving, finding, and calling a particular number. The team also wanted to make selection adjustments brainless, like the mandatory manual network selection imposed by the GSM standard. While going through the task scenarios, it became obvious that it would be impossible to meet the GSM standard without a menu. Consequently, the new UI style for a consumer phone had to provide call handling as well as menu browsing and selection.

The design team ended up presenting a concept with three main control keys: a dynamic softkey, an arrow key, and a clear key.

Back then two softkeys were regarded as one too many, particularly if one does nothing more than backstepping. A single-softkey solution was visually clear and easy to localize. The key could be centered, and the user did not have to make a binary choice. The single softkey was such a graceful solution, and really focused on the user's primary usage path. (For more about the design challenge softkeys represent, see Chap. 2.)

An arrow key seemed like a natural complement to the softkey. Consumers knew arrows from several other devices such as remote controls. We went so far in reducing the key count that we removed the up arrow as a means of scrolling backward. Because the menus were closed loops having only few items, it was possible to navigate a menu by one way scrolling. Backward scrolling was regarded as redundant functionality. The only arrow was pointing down, as we expected most people to have a mental model of scrolling down into a menu.

The C key (C for clearing) was also considered intuitive. It was very well known from previous Nokia phones and from all calculators. Only later did we discover a drawback—because our concept lacked a SEND key, users also found it intuitive to understand C as call.

In addition to the three main control keys, the concept had a separate power key and the standard 12-key numeric keypad. A phone that is not easy to turn on and off is flawed even before use, and thus we considered the power key an obvious prerequisite. Being able to turn the phone off quickly and without fumbling is important for polite phone manners and for privacy protection.

After some iteration with menu structures, we were ready to have the first PC-based simulation programmed using one softkey, a clear key, and a down arrow. We did the simulation at Nokia Research Center's usability lab and immediately learned two things:

- Users often scrolled past the desired menu option and needed to step back. Scrolling through the whole menu was a far too clumsy way to correct a minor and frequent error.
- A single arrow key down offered no mental model for navigation in a menu. The users didn't associate it with scrolling the items presented on the screen.

Figure 3.1   Screenshot picture of the first Navi-key simulation.

So we added an up arrow and ended up with four main control keys. Fortunately, no other major problems were found, and Navi-key user interface (Fig. 3.1) was born.

Launching the Concept

The design team realized that getting the concept accepted would be the hardest part of implementation. Before the new UI could be put into a product, it had to be accepted by management and by the product implementation team. We decided to present it to management first, which seemed like an easier route than convincing the notoriously tough project managers.

It was difficult to get enough demo time to convince management by means of a computer simulation. Simulations are cumbersome, removed from the real context, and often inconveniently incomplete. What we really wanted to do was to make an unofficial pitch, inspired by the David concept that had been demonstrated in a real working phone with great success. Mikko With, a brilliant young engineer, was assigned to hack the software as a skunk work project sponsored by Juhani Miettunen, a veteran in the UI software community. Mikko was barely out of university, but very sharp and keen, and we had the first functional demo running in about a week. It was fantastic to see everything come together, even if the first demo software had bugs and some mistakes were evident. We tweaked the phone for another week, with the author specifying tweaks via phone while commuting between Salo and Helsinki and Mikko while working from Oulu in northern Finland. One of the things that we removed from the software was a menu introduction screen probing users to scroll. This seemed to confuse users as they thought the same screen appeared regardless of their selections. This intro screen has never since appeared in any of our phones.

The hardware prototype was built by modifying previous Nokia 2148 phones (Fig. 3.2). New plastic covers were made up and brightly painted for it. The key layout in the prototype was strange, because we had removed several keys from Nokia 2148, leaving an empty space in the middle of the front cover. As industrial design wasn't an issue, we were worried about the phone being ugly, but we later learned that the lively blue prototype was so polished that people actually mistook it for a real phone. The appearance of the key layout got lambasted as a consequence; fortunately this critique was easy to explain away. We made about a dozen prototypes, which were handed out for use by senior managers and other people with influence in marketing and R&D.

The mentors, Erik Anderson and Mikko Palatsi, were both very excited about the concept and did a lot of pitching behind the scenes. One of the most important targets was to convince the top management. To our glee, we didn't require any convincing beyond the prototype and immediately we found a champion for Navi-key in management. One of the board members described Navi-key as the future of mobile telephony and exactly

Figure 3.2 Hardware prototype of Navi-key UI style made by modifying Nokia 2148 phones.

what Nokia needed to make phones penetrate the broad consumer market. Without this support there would have been only a dozen handmade phones. The scenario is reminiscent of the story of the birth of the Walkman, a controversial product for a television manufacturer until a visionary named Akiro Morita became its champion. Our belief is that product champions are a necessity for getting radical innovations to the market.

Søren Jenry Petersen stepped into the role of program manager for the initial implementation of the Navi-key user interface in the Nokia 3110 phone (Fig. 3.3). His first words to the author were "So, here is the man who is screwing up the schedule of my project!" Søren is notoriously known for a flamboyant way of expressing himself. This encounter could have turned out to be a disaster if we hadn't understood each other. However, Søren became convinced that the new UI style would be a real differentiator for his product and provide a great opportunity for the newly acquired Danish R&D site. He championed it heavily in Copenhagen, discerning in Navi-key Copenhagen's first opportunity to put its mark on Nokia products. After the 3110, Nokia made several Navi-key phones, each one outselling the previous model.

Next the Danish engineers had to be convinced that it could be done. Fortunately the team in Copenhagen consisted of experienced former Philips engineers who had made mobile phones before. However, the Nokia software architecture and procedures were new to them. The chal-

Figure 3.3 The first Navi-key phone, Nokia 3110, was introduced in 1996.

lenge from management was that the Nokia 3110 phone incorporating Navi-key had to be launched in March 1996 at Cebit and shipped 6 weeks later. That stipulation left about 11 months, and in retrospect meant working at rocket speed. Many of the old-timers were skeptical that it could be done, but Danish Viking power seemed to take over.

Product Development and Testing

The next few months were hectic. The concept needed polishing, UI bugs had to be ironed out, the concept had to be verified with users, and specifications had to be written for implementation. Finally it had to be programmed and tested. The software group led by Lars Bergmann and Flemming Klovborg made a tremendous effort. Flemming and his team wrote the specification in the record time of 8 weeks. The implementation path was to modify the existing software from Nokia 8110. Such a leap could not have been made if the phone had not been based on existing software. Fortunately, we had a very flexible architecture well suited to style variation. Within 5 months from project start, the first prototypes were on the bench.

Seeing the Navi-key user interface style live and working in a phone was almost like seeing new life. It was a major energy booster. The phase after specifications are frozen and the first software builds are released is a dark time of skepticism, but when the first build runs in the prototype hardware, everyone becomes intoxicated with energy.

The final detail we needed to solve was the design of shortcuts to phonebook and the redial list of recently called numbers. After considerable debate, we decided to depart from the logic used in other Nokia phones, where scrolling down from the idle screen discloses the first name in the phonebook in alphabetical order and scrolling up discloses the last one. In Navi-key UI scrolling down takes the user to phonebook as usual, but scrolling up opened the redial list. The pair of up and down arrow keys that were so visually connected that they seemed to consist of a single scroll element actually provided two different functions. The solution was not consistent with those for other phone models and was rather counterintuitive, but this was a case of the art of logic. Simply put, redial was too good a feature to remove, and with the few keys available in Navi-key UI no other options were feasible. (See Chap. 2 for the rationale behind shortcuts.)

While product development was ongoing, we still harbored uncertainties about the overall usability and acceptability of the Navi-key concept. We did not want to fall into the "bimbo trap" again and launch a phone that looks so easy that it's insulting. Another lurking trap was the "marketing ease-of-use trap," where a product at first glance looks like a snap to use in the store and then becomes a bear to operate when you power it up at home. We also had concerns about whether softkey calling without a dedicated SEND key really worked. The first concept test clearly showed that about 30 percent of users made error calls with their phones during the initial phases of using Navi-key. This figure was a bit higher than for phones with SEND and END keys. And finally, some users would lack the slightest idea of how to interpret a softkey: absent that understanding, one is pretty much lost with the Navi-key. There were numerous examples of people saying "Oh, you work for Nokia, I really like your phones, but what are the texts in the lower part of the screen used for?" We even encountered two people who, after a 30-minute test, still did not have a clue of how the phone worked.

We decided to test the possible obstacles for Navi-key UI thoroughly. We needed to create a more conventional solution for a benchmark, so we mocked up a scrolling Ringo (as it sounds, a Ringo phone with SEND and END keys and up and down arrows). Our tests were performed with real prototype hardware. This was a unique case, as most design is done using PC simulators or paper prototypes. In the comparative test of Navi-key UI and scrolling Ringo, we noticed that using the Navi-key phone was regarded as a more positive experience than using the scrolling Ringo. At first users were skeptical about Navi-key, but the concept's simple mental model stimulated learning. Skepticism turned into learning and even approval. Maybe we had created user delight: a positive experience beyond expectations—the ultimate goal in all our work.

While the scientists were doing solid research work, team manager Lindholm was doing some impromptu fieldwork of his own. He conducted more than 20 parallel informal usability experiments to tax the concept to be able to defend it to the inevitable raft of internal skeptics. It was necessary to learn how users reacted to calling with softkeys instead of dedicated SEND and END keys. Lindholm was obsessed with the rate of error calls and how quickly people realized their mistake. He selected complete cellular novices for the test subjects to see how they learned. In the mid 1990s it was still possible to find people in the company who had never made calls on mobiles. Nokia was growing quickly, and not all people joining the company were telecommunication engineers. Some outsiders with no connections to Nokia were also recruited. About 50 percent of these people called by mistake when pressing the softkey, but most of them made only one error call. In fact, most of them were able to correct their error before the call was connected.

The most decisive of acceptability tests is the "spouse test." The rationale is that a spouse would never give praise for the sake of praise. The author is fortunate to have a wife who is not technologically driven and who represents what could be considered to be a normal person. She breezed through all the tasks with great relish, which was a surprise as she had previously expressed no interest in mobile phones. She has been a sworn Navi-key user since the product was launched.

Knowing the users' first impressions was important, but not enough. We decided that we wanted to gather more data on how their perceptions of the interface evolved over prolonged use. Accordingly, we arranged a test where six people received Navi-key phones to use for 6 weeks. We gave them only the most basic instructions about the user interface to start with, but we also set up a help desk that they could call. The 24-hour help desk consisted of a single person—Miika Silfverberg, a psychologist and an expert in mobile user interfaces. He promised that the users could call any time, day, or night, and clearly he was betting on getting his sleep. During the entire 6 weeks, Miika got one call. The users weren't slacking; they were just happy, and they had used most of the implemented features.

These latest tests were done partly to find confirming evidence for the concept, as we had started to run a new project (a follow-up basic category phone, Nokia 5110) for which the Navi-key user interface was under consideration. The long-term test, of course, could not have affected the Nokia 3110, as we already were past the point of no return. We were in fact in the predictable "generation trap," where some critical parts of the next generation had to be frozen before we could collect real feedback from the previous one. This dilemma between continuity and evolution is difficult to manage in new-product creation, and unfortunately it occurs constantly. The way to solve it is to ensure that people are motivated enough to do several consecutive projects, where learning can be passed from one to another. Nothing beats the tacit knowledge of past experience.

By now we had performed lots of different tests on the phone and used this phone ourselves daily. We were feeling quite confident about its usability. What we didn't know anything about was the consumers' willingness to buy it. Generally we avoid the temptation to think that a user interface alone can sell a product. Usually the physical device characteristics, such as appearance, shape, and weight, have a more powerful impact. We therefore decided to wait for prototypes and then do a comparative preference test to acquire the statistically significant data we needed to galvanize our own salesforce.

Marketing the Product

Our salesforce was used to the Nokia 2110 and they liked it; the product was easy to use and sold in great volumes. So there were yet more people to convince of the superiority of Navi-key UI over previous Nokia user interfaces. We wanted to demonstrate that Navi-key was better not only than Nokia 2110 but also than the competing products. We concocted a test to be conducted in a London train station, where we randomly selected people carrying mobile phones and asked them to perform the tasks of saving, finding, calling, and deleting a name and number. We used the Nokia 2110, the participants' own phones, and the Navi-key prototype (Fig. 3.4). The Navi-key phone turned out to be the easiest to use, even easier than the phones people were accustomed to using. This result was so encouraging that it almost seemed fabricated—it was not. We genuinely had managed to make an easy-to-use digital mobile phone. The results were never used in external marketing.

Eventually the Navi-key concept was selected for the Nokia 5110, which was launched a year after the Nokia 3110. Second-generation

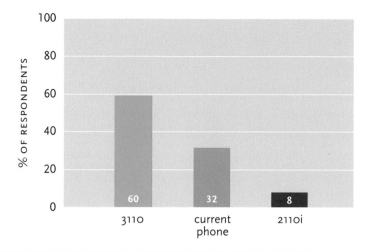

Figure 3.4 *The test comparing the Navi-key UI 3110, the 2110i benchmark for ease of use, and the user's own current phone. The 3110 Navi-key phone scored as a clear preference in terms of overall ease of use.*

Navi-key UI was different from its predecessor in one respect—the C key was now a backstep key, taking the user one level back in the menu, instead of an exit key taking the user to idle mode. We thought that a user interface without a panic key in the form of EXIT is flawed, and we had endless discussions about this issue. In the end we discovered that novices indeed preferred the EXIT option, but only for a little while. Once they became more familiar with the phone, they started to value the ability to skip one level in the menu. We learned that a user interface has to have a logical way out, but not necessarily a panic exit.

The hardest things to do with a mobile phone user interface include multiparty calling, call waiting, and other advanced in-call operations. The enthusiast could experiment with the different Navi-key UI versions—Nokia models 3110, 5110, and 3210—to find that they are all different. Handling two calls simultaneously means that you need three keys to be available on equal priority. You need to be able to end the current call and answer or reject the waiting call. When you have two simultaneous calls—one active and one on hold—you need to be able to end the active call and switch to the call on hold. Initially we decided not to implement this supplementary feature at all, because we deemed it too difficult to use with the simple Navi-key UI. The nonsolution, however, proved to be a nonoption, as Nokia's operator customers clamored for multicall handling because it is a great source of revenue. Usability problems were not, however, sufficient to discourage us from implementing this capability.

The use case we optimized for was called "the human answering machine." This means that the existing conversation has priority, but it can be interrupted for answering the second call. The user is offered an "ANSWER" option on the softkey. On pressing the key, the first call is put on hold, and the user can give a personal live greeting to the second caller, such as: "Hi, I will be on a call for 15 minutes; I will call you back." On the second press of the softkey the user would end the incoming call and return to the first call. This worked very well in Europe, but in the United States, users prefer that neither call be prioritized. United States users insist on the ability to continue either call. This forced us into having "OPTIONS" as the softkey command during calls, which made the entire process one layer more complicated. If we had known how important this feature would be in the United States and how much hassle it would cause us, we might have trashed the entire Navi-key concept.

Why has Navi-key been so successful? There are five unique qualities in it.

1. Its *key count* is lower than the competition's, even though the difference is only two keys compared to Nokia Series 30. Apart from the number of keys, their nature and their logical positioning on the phone create a distinctively different keypad, which people seem to notice.

2. The initial *perceived ease of use* really delivers on its promise. Its thoroughgoing consistency stimulates learning. At first you learn one or two features by rote, and after a while you can deduce how the rest works yourself.

3. The Navi-key user interface has proved to be very *flexible for adding new features* without hitting the usability knee (see Chap. 1).

4. We have noticed that it is difficult to create bad *feature design* for Navi-key. As complexity increases, the way to design for Navi-key is to put the new commands into an options list and then sort the list into a meaningful order.

5. Finally, the *softkey* itself, with its printed blue line and changing screen label, adds an element of mystery. The users have described the key as being intelligent: "It somehow guesses my next actions"; "It tells me what to do next"; and so on. Our marketing people seized on this element of mystery and dubbed it the "Navi-key." This name has since become synonymous with ease of use, and other Nokia products have leveraged it as a usability trademark. Some people responsible for Nokia brand have claimed that the Navi-key is one of the most valuable subbrands we have.

Navi-key has surprised us time after time with its versatility. It has managed to evolve as new features have been added. Many thought it would be a disaster for browsing, but our experience was that a single softkey serving an options menu in a third-party service actually tends to harmonize the services and lowers the barrier for usage. Obviously efficiency will never equal a dedicated select key concept, but it holds up as acceptable. The key problem with browsing is not the keypad, but the size of the display.

Today the biggest threat to the Navi-key style is the number—not the type—of features that have to be implemented. It will be very interesting to see how the industry evolves: whether there will continue to be a market for a good plain old telephone or whether they will be replaced with sophisticated browser phones. Our bet is that the market will continue to segment and customers will acquire a more nuanced concept of "phone" by stages. A phone is no longer a phone, just as there is no longer a screwdriver, but only a multitude of different tools with slightly different objectives.

Conclusion

We believe that the Navi-key logic, because of its versatility, would be very suitable for use in other electronic products as well. We have played around with using the concept in wrist wearables, and as long as there is room for at least a two-line display—one for content and one for command—it works just fine. Consider the example of cars, which are undergoing an information technology (IT) revolution. The dashboard has traditionally been a collection of user interfaces for heating, radio, lights, and other components, and the design philosophy has been that each function has a dedicated knob, or lever. With increasing functionalities, this philosophy leads to the airplane cockpit syndrome, which is probably not the desired scenario. The Navi-key logic could be used for many functions in a car. Because it is compact, it would not clutter up the instrument panel or even the steering wheel. And where else can you find a UI concept known by 300 million users?

This story could have been taken out of an innovation textbook where novices with their fresh insight and courage challenge the conventions. The Navi-key design team suggested demolishing the pillars of the successful Nokia 2110 user interface. The story also featured mentors who spent and exerted much time and effort without claiming credit. It also has an element of skunk work with the creation of the prototype, which finally sold the concept to the product champion in management. Finally it featured all the engineers who actually developed the phone and the software. This was the product development fairytale. Future advances can be made only by rushing back to the lab!

PART 2

Living with Mobiles

Mass manufacturing organizations tend to be fragmented by functions and by the professional backgrounds of people working in them. Each fragment has its own kind of user orientation. Ergonomics, marketing research, quality engineering, and other disciplines have long traditions of locating a user, using all their gauges on that individual, and then pushing their findings into the design fray. For user interface designers, user orientation is usability. *Usability* has been defined in a very broad and inclusive manner as "the quality of use in context."[1] However, the practice has focused heavily on task-centered thinking. If a given user accomplishes a given task quickly and without mistakes, the product is usable. Understanding the user is in effect understanding how that person performs the relevant tasks. What particularly characterizes the discipline is just how detailed this understanding has to be. The tasks to be evaluated are deconstructed into the smallest pieces imaginable.

The other end of the spectrum—where the user, the product, and the usage itself are construed as parts in a bigger picture—has been less usability-engineered. However, that is exactly where we are starting to find our biggest usability problems. The basic solutions of UI styles and processes have matured. We know our menus and softkeys pretty well, but we constantly face problems in understanding why people browse those menus, or why they don't. The "why" is, of course, a user interface design issue, but the answer lies beyond the reach of cognitive *task centered usability* and the associated user conception. Thus, we have identified a need to widen the scope of designers' user understanding. There are several reasons why a broader take on user motivations has become increasingly relevant, especially for the mobile communication industry.

The obvious reason for breaking through the inherent limits of task-centered usability is underlined in each and every paper written about mobile user interface design; the user has stepped out from a lonely control room or an air-conditioned office into the public space where she (or he) wanders around the physical world and encounters the people living in it. And that really makes a difference. In a bus, in bed, on the beach, while baking with her hands covered in dough, when riding a bike or driving a car, alone, in the midst of an intimate discussion, among customers, in a snow-storm, when still very young, after she's grown old—the list never ends. All these conditions have design implications calling for *contextual knowledge* about the user and her communications.

Mobile phones have already been domesticated, that is, have become an integral part of the consumers' everyday life. Using the technology cannot be separated from living the rest of their lives and put under a microscope like an isolated object. The ubiquitous presence of communication technology makes it more appropriate to think of people as relating to mobiles than as operating them.

When observing people relating to an artifact, task centrality stops being the only or even the dominant point of view explaining the relationship. For clothes, personality and fashionableness go with—or before—practicality. Houses are not primarily efficient living machines, but cozy and intimate environments. Watches are jewelry that happen to tell time; cars are lifestyle in addition to transportation. Correspondingly, living with mobile communication technology introduces a diversity of subjective, emotive, and expressive criteria. Users have to feel that their lifestyles and preferences are in harmony with the products they keep around. The user needs to be seen as a subject *consuming technologies.*

Mobile phones and mobile communication technology have been hyped a lot. It has been easy to get the idea that 3G, WAP, Bluetooth, and the rest will change the world overnight. The actual solutions implemented have improved a lot and given the user real value, but (no surprise here) not at the pace of expectations. Consumers' expectations toward mobile communication technologies are constructed from their own exposures, past experience, and future promises. There is the users' firsthand experience with 2-year-old cell phones having none of the features hyped. There is a lot of excited talk about mobile Internet, natural-language speech recognition, third-generation networks, and fast wireless connections. There is a tradition of computing associated with respect, skill, and professionalism, and another associated with plug-and-play, idiotproof consumer durables. There are techies crazy about all novelties, and luddites who insist on hating it all. A consumer cannot evaluate a product and its features solely on the basis of practical utility. New solutions are introduced with fancy names and seductive descriptions. How to position a new feature? Which tradition would be the most appropriate reference for fostering understanding of a product? With which tradition do I identify myself? Products launched by the mobile industry are now positioned into a structure of values and references. *The socially constructed meaning* of a product influences or even determines how it will be accepted by the audience.

Mobile phones are tools and toys. If I call because I need a taxi, the phone is an instrument to get something done; my interaction is extrinsically motivated. If I write a text message to my friend, because I want to think about her for a while, to devote some of my time to her, the writing is intrinsically motivated. I don't mind that the correct punctuation is difficult to enter, as I enjoy polishing the message. I play snake because I am hooked. There is no need for a motivation of any kind—interaction as such is so engaging that it keeps me pressing the keys. The very same product, even the very same feature in it, can be used in different situations for very different reasons. We need to understand the range of motivations a user may have. Our image of the user needs to be widened to include motives, customs, and apparently even addictions.

Communication is a prerequisite for social behavior. Without communication there would be no societies, no cultures. Spoken face to face, language is the primary means of human interaction. Then there are artifacts based on language, and capable of enhancing the power of language. When we design these tools, we come close to the basic principles of how societies and cultures function. Designing for our compatriots is efficient because we have a complete understanding of the cultural codes without taking any specific actions. We can just start focusing on the task-specific issues. Designing for people who seem exotic to us is hard because we cannot take any codes or interpretation frameworks for granted. We need to step back and look at the whole culture to be able to evaluate the relevance of even our task-level questions, much less the right solutions. *Cultural end-user studies* are the way to attain global user interface design.

Besides the discipline-specific challenges described above, there is, of course, the generic core design challenge—designers of new technologies have never seen the world for which they are designing. The very existence of novel technologies changes the situation in which they are used, and the use changes the products. So understanding users means understanding how they change as the society around them changes in general, and specifically how they change through interaction with the products that we introduce.

To sum up, a mobile phone user is

- An information processing unit accomplishing tasks
- An actor in varying physical and social contexts
- A consumer with a lifestyle
- An interpreter of socially constructed meanings
- A locus of different motivations
- A member of a culture
- An object and an initiator of continuous change

Obviously discussing user needs—and we still use the word "need"—is no longer solely a matter of addressing the practical and functional demands related to accomplishing a specific task. Nor do we use the word "need" to refer to the difference between a subject's present state of being and the target state, as some marketing theories do. For us, user need is any relationship between a person and her context that may have an influence on the design of products and services. The relationship may be physical, behavioral, motivational, or driven by values, interpretations, and cultural codes.

We can't say that we understand our users perfectly. That would be impossible. We have, however, recognized the imperative for comprehensive user understanding, and taken steps to improve our sensitivity to end-user needs. Organizational changes have been made to encourage cooperation between marketing research, usability research, contextual user studies, and sociological consumer analysis.

In Chap. 5, Riitta Nieminen-Sundell and Kaisa Väänänen-Vainio-Mattila describe the link between communication technologies, the social change they are engendering, and the new research and development approaches that change demands, discussing the manner in which sociological research tradition can equip product development organizations with appropriate tools.

We have also gathered plentiful experience about contextual design. A case study from India by Katja Konkka (Chap. 4) scopes out the results that can be collected by such cultural end-user research. It is astonishing to see how different kinds of issues which all are relevant to mobile communication can be exposed by a single research approach.

Katja Konkka

CHAPTER 4

Indian Needs
Cultural End-User Research in Mombai

Visiting a typical Indian family in a suburb of Mombai (formerly known as Bombay), or spending the morning with an Indian journalist at his or her office? Perhaps following a businessperson to a client meeting in New Delhi? This is a part of your work if you throw yourself into cultural end-user research. Well, of course you also read books about different countries, cultures, and religions. But in order to feel, hear, smell and see the actual context for which you're designing, and to understand the daily life of your customers, you must experience the environment and meet the people.

Culturally Varying Needs

We talk about functional and emotional end-user needs. By *functional* end-user needs we mean that new designs have to serve the practical objectives of the users. They have to support existing practices, workflows, and organizational roles. *Emotional* user needs determine the user's pleasure when using technology.[1] Technology can be pleasurable if it gives the user a sense of control, or if its use attests to a high degree of skill. Pleasure can also be evoked by aesthetic appearance or positive associations that attach to a technical object.

The word *need,* as it is used in this chapter, probably reflects everyday language at Nokia more than any formal definitions. Our concept of needs can include environmental settings, cultural values, economic possibilities, legislative regulations, religious beliefs, or language-related interpretations—whatever issues are observed to have a potential influence on mobile communication and the design of communication products. These needs are related to user interface solutions either directly or by providing a context for interpretation.

Functional needs vary culturally, and emotional needs even more so. What is pleasing for the eye or ear, what is socially acceptable, or what evokes positive associations are indeed cultural matters. Colors, for instance, have culture-dependent meanings. Some South African tribes see red as a friendly color and green as a hostile one. These people may have problems interpreting traffic lights when visiting urban areas. Some Indians see orange as associated with Hinduism and green with Islam. The interpretation of icons and graphics can be very culture-dependent; for example, a Hindu sign meaning "all the good" is often perceived by Westerners as a Nazi swastika.

The vast potential of mobile market opportunities can be wasted if we are not prepared for each different culture entering the mobile information society. Preparation essentially consists of working toward an understanding of the everyday needs of those populations. If we don't understand our markets well enough from the perspective of end-user needs, the new features we create will not be accepted and—even worse—we won't necessarily know why.

Responding to culturally varying end-user needs doesn't always have to mean completely different products for each market area. Quite small user interface (UI) issues can sometimes make the difference. For example, the user's ability to add Hindi greetings or religious symbols to a message adds a great deal of emotional value to short message service (SMS) in India among its Hindu people. Hindi music in ringing tones and other alerts would do the same. The scalability and variability of the whole product to suit different cultures is a basic requirement for cultural adjustment.[2]

We put a great deal of effort into knowing our customers. Understanding is acquired by conducting different types of research projects in various cultural areas—projects ranging from traditional market and usability research to contextual ethnographic studies. During the year 2000 we conducted an end-user needs study in India.

Apprentice

There were approximately 1.95 million mobile phone subscribers in India as of April 2000. The number of subscribers was tapped for constant growth. Although the penetration percent at the time—0.5 percent in early 2001—was low, note that 1 percent of population means 10 million people in India, which has about 1000 million citizens. Thus, even a small percentage of growth would be meaningful from the business point of view. Certainly, we recognized that the majority of Indians will not enter the information society in the short run. However, it is reasonable to believe that the next 2 to 5 percent of Indians will reach the living standard of current mobile phone users in the near future.

Government regulations and tax policy previously slowed the growth of the mobile phone business in India. Current regulations make the mobile phone owner responsible for charges from incoming as well as outgoing calls. This is expected to change in the near future. As a consequence, the number of company phones will probably increase, because employers will be able to control the bills better. Also, private use of mobile phones is expected to grow when "calling party pays" (CPP) charging is introduced. At the moment mobile calls are 5 to 6 times more expensive than landline calls.

The gray market for mobile phones is strong in India. Phones appear in the gray markets months before they're officially sold by outlets and operators. In India, operators regularly receive queries about phones that are not yet launched. Prices in the gray market are cheap—approximately 50 to 60 percent of the official prices—so official dealers are hoping for reductions to the 40 percent taxes set for imported goods.

We had lots of numbers like this. The numbers can be used to construct a macro image of India as a mobile phone market, but they were far from sufficient to guide the design and localization of products for end users. Our project aimed at going from numerical abstractions to the concrete levels of users' lives. We were there to explore culture-specific factors, which could potentially affect user interface design. The study consisted of three phases: analysis of background information, an interview study in New Delhi, and, as the primary objective, user observations in Mombai. We selected a qualitative research method for the project because of our previous experience with the utility of ethnographic research methods. Our end-user needs team has a strong expertise in them, and we trusted the approach.

The shift from place and time dependence to anytime/anywhere information access has set new requirements for developing personal devices.[4] We have to understand the users' physical, social, and cultural environments and how they are affected by other users' activities, roles, and values. It is difficult to acquire this kind of understanding with quantitative market or usability research. Not even the questions are known when we begin. There are no hypotheses, and only some focus areas of interest can be defined. From these vague starting points, ethnographic research methods such as *contextual inquiry*[3] provide a structured way to gather contextual information on users' activities and related cultural aspects.

The core research team consisted of linguist Johanna Tiitola, usability specialist Dr. Minna Mäkäräinen, and the author, a psychologist. In India,

Indian cities are noisy and crowded; you can hear traffic noise, car horns, and people talking, shouting, and selling everyday goods. People cover one of their ears with a hand so that they can hear when talking on a phone. Additionally, people need to use a hand in front of the mouth to cover the space between the mouth and the phone to protect from background noise.

local usability experts and translators joined the team. A multidisciplinary team is necessary in this type of research. Experts with different backgrounds and skill sets tend to focus on slightly different things, which helps in building a holistic and comprehensive view of the research topic. When design implications are drawn from the collected and analyzed data, a multidisciplinary team can consider solutions that might be implemented in different phases of product development.

Using local help is absolutely crucial when conducting cultural research, not only to translate between languages but also between cultures. We involved local human–computer interaction specialists with expertise in the field of mobile communications. However, company-internal employees are needed for conducting the research because they know their own company and its business objectives and products. Without such understanding it is difficult to sharpen focus because—as mentioned above—needs are product-related matters.

The research team was aware of the richness of India as a cultural region when choosing the two cities to be visited. New Delhi and Mombai were selected because both cities have a developed mobile communication culture on an Indian scale, and they can be seen as trendsetters for the rest of the country. Still, it has to be remembered that the study's results are oriented to New Delhi and Mumbai. India is a vast area, and one needs to be careful with any generalizations.

We followed the basic rules of the contextual inquiry method:[3]

- Go where the user is and see an activity as it occurs.
- Cooperate with the user to comprehend his or her activities.
- Make interpretations to ferret out the design implications of the users' behavior.
- Focus on the activities most relevant to the design.

In addition, we collected artifacts, that is, objects of interest telling about communication and personalization needs. Conducting this kind of user observation requires an open-minded attitude and adequate social skills. A researcher has to entertain new ideas, be communicative in turn, and willingly learn from the users who the experts are when it comes to their everyday lives. The researcher has to become the user's apprentice.

The Indian Way

Conducting observations in India was a pleasant surprise for our Western minds. People were friendly and had a positive attitude toward our study; they had no problems inviting us to their homes and offices. During the study, all family members were present and actively participating. The families seemed to have a joint experience of mobile phone usage rather than strong individual opinions. It became obvious that experiences with mobile communication tools are shared between friends and family members.

Some informants found it difficult to express negative experiences with the devices they knew to be designed by the company we represented; we had to make it very clear repeatedly that we want to improve our designs, and in order to do so must know what could be done better.

Typical users in the Indian mobile phone environment are upper-middle-class and wealthy males. There is still excitement about mobile phones; mostly businessmen and executives use them, although nonbusiness usage is increasing. Possession of a mobile phone is significantly related to higher education, English speaking, and the use of high technology. Thus the current mobile phone users are the ones who are already comfortable with technology. (This won't be true in the future.) Status attaches to the mobile phone model the user carries. However, just owning a mobile phone is a status symbol, and is related to a high standard of living.

Apart from these obvious phenomena often found in the early penetration phases on new technologies, there are several more interesting and more native phenomena in Indian mobile communication culture that clearly distinguish Indian mobile phone use from Western mobile phone use. The results below explain the context in which mobile communication transpires in Mombai and New Delhi. They also illustrate the scope of issues that cultural end-user research can cover.

Noise

Indian cities are very noisy. Traffic, chatter, yelling, hawking, and car horns make the environment difficult for phone usage. People shout into the phone, cover their other ear with their free hand to block out noise, and cover the space between mouth and phone to protect the microphone from background noise. Ringing tones successfully tested on the main street of Tampere, a Finnish town with 200,000 inhabitants, would not be noticed in India. Effective background noise reduction and alerts for noisy environments are needed.

Environment of Extremes

Hot weather, monsoons, humid conditions, dust, and bright sunlight followed rapidly by dark and unlit spaces—all of it affects phone usage. The phone should be moisture- and dust-resistant, and the screen should be viewable in both very bright light and near-darkness. This makes a designer think twice about the usability of, say, a touch screen in India. The practice of eating with one's hands doesn't make it any easier.

In India amulets are commonly used. This amulet is made of lemon and chilies, and it is used for expelling bad spirits.

Long-Term Investment

In India a mobile phone is a big investment, even for upper-middle-class people. Consumers carefully consider which mobile phone to buy. Materials are reused as much as possible, repairing things is preferred to buying new, and technical purchases are regarded as long-term investments. Phone lifecycle is expected to be long. One criterion an Indian mobile phone buyer might consider is whether today's phone purchase will be able to surf tomorrow's mobile Internet services and sites. Thus, spare parts and aftersale services are important, as are feature updates to keep the phone modern throughout its lifetime.

Guessing the Caller

Indians do not introduce themselves when answering an incoming call. Well, not all Westerners do, either. But Indians don't just begin with "Hello!" or "Pronto!" They immediately start asking "Who is it?" or guessing "It must be my uncle who is calling." One should be able to recognize a caller by his or her voice. This applies mainly for friends and relatives, but also for regular business partners. To guess incorrectly is considered rude. *Calling-line identification* (CLI), which is already a standard mobile phone feature, indicates the origin of the call by displaying the number or saved name on the handset screen. It becomes much easier to avoid being rude with CLI—unless someone other than the phone's owner happens to call. But, we have to wonder, does CLI eliminate more pleasure than it creates? Recognizing someone by voice is an indication of intimacy, devotion, and skill. The caller probably perceives an essential difference between being recognized by his or her voice and being recognized by his or her subscription ID. Has our adaptation to communication technologies suppressed our sensitivity to these kinds of issues?

Social Event

In India a phone call is a social event. When one family phones another family, such as relatives or friends, each family member is expected to talk to each member of the other family. Conference calling is not accepted for this purpose because it is not considered personal enough. It is used only for saving money, such as when calling overseas. In Western countries, conference calls are fully acceptable even when communicating with family members. So how could UI design make the conference call feel personal enough for an Indian phone user? For instance, could there be a readout identifying who of the conference call participants is speaking at the moment?

Village Connectivity

Housewives in some parts of India and Bangladesh keep village mobile phones, and rent airtime to other community members. This so-called village connectivity is a fair start for introducing the mobile phone culture into the smallest villages as well. A big issue when moving from landline to mobile phones has been the shift from the mindset of calling a place to calling a person. Much mobile phone design philosophy has been related to this change. We have been pushing the personality and individuality of such communication as far as possible. And this direction has been accepted by Western markets with enthusiasm. However, "village connectivity" places very different demands on phone design. To whom is an incoming call or message addressed? How will costs be shared? Which entries in the contact directory are mine? Are these questions formulated so much from a Western individual's viewpoint that they are completely irrelevant? Will mobile phones start to change social behavior towards more individualism in India also?

Sharing

The sharing of products and information is very common in India; phone calls can be shared with family members, friends, and colleagues. Emails can be printed out and circulated for comment, and they are usually sent to groups of people rather than to one person. It is actually considered unfriendly to omit someone among friends or colleagues from the distribution. On the home front, it is popular to have a common telephone directory or to share a mobile phone with family members and friends; for example, sons and daughters ask to borrow dad's mobile when going out.

Even more interesting is the practice of group Internet surfing. The opinions and experiences of friends and family members are very important when selecting a device to purchase. Discovering something in common, such as smoking or having the same model of mobile phone, even connects people who don't know each other. Group communication, trust, and emotional bonding are highly valued. What are the design implications of sharing? Should we provide personal profiles for users sharing a phone? Would Indians use a group SMS and chat for group communication? Would they benefit from an option to share the same view on several screens when browsing the Internet?

This is a gift to Lord Ganesha. Lord Ganesha, perhaps the most loved deity in Hinduism, is a symbol of knowledge and "auspicious beginnings." He is prayed to before the beginning of venture to ask for his blessings.

Time

A calendar application on a personal digital assistant (PDA) must function as a time management tool supporting the models for planning, scheduling, and making appointments that the users know and like. In India, time management is usually based on daily, weekly, and monthly routines. Calendars are used mainly for checking the dates, not as reminders, as is the case in Western business time management. Appointments are often memorized, and post-it notes are used instead of calendars as reminders. Appointments in the far future are tentative by nature until an exact time and place are set closer to the event. Thus, a functional calendar application for the Indian PDA markets is optimized for routine-based time management, and calendars present entries in the form of notes and to-do lists rather than reminders. Perhaps calendars should be optimized for presenting dates and weekdays, and there should be a separate reminder application utilizing a post-it note metaphor. Should there be a specific UI solution for tentative appointments? Maybe the act of creating a note or reminder about an appointment already verifies it in such a way that the previous model of tentative meetings would be destroyed. It is not always possible to transfer existing patterns of behavior to information technology applications without changing the nature of the task.

Orientation

In India, people rely on asking directions and spotting landmarks to find places instead of using maps. Such restaurants, hospitals, railway stations, movie halls, government buildings, religious monuments, and other sites or buildings are popular landmarks. People consider them to be more stable than streets. For instance, a meeting place is not defined by address, but by giving directions and describing landmarks. UI design can support this by allowing fields for entering places in contact applications, which aren't restricted to address, and adding a space to explain directions. By the same token, a travel guidance application for the Indian user should not rely on street names, but give the coordinates for direction and distance, and include landmarks. Navigation applications based on landmarks and directions instead of streets will need a

completely new interface. How might that influence the design of location-based services?

Mixed Language

Our research project had a secondary objective to collect opinions about
Nokia 3210, which contains the world's first Hindi UI. We found that the
technical terminology in our first Hindi release was rendered in literary
and somewhat old-fashioned language. The results revealed that many
technical terms would have been more understandable in English written
with the Hindi (Devanagari) script. For instance, the terms *call divert,
data, fax, call register, prepaid credit, incoming call alert,* and *infrared*
don't have good Hindi analogs. In spoken language Hindi and English
are mixed, and sometimes Roman script is used to write Hindi. An example of a "hinglish" sentence is "tu tension mat le" meaning "don't get
tense." English terms are used extensively, especially in technical speech.
We have done our best to make the UI as localized as possible, but
haven't paid enough attention to the fact that much of the language is in
fact localized as English. We still don't know, however, if what we did is
right or wrong. Localization is not that straightforward because understandability is not the only criterion. Native language can be important as
such, and using local languages shows respect for the local culture.
Applying expressions in Hindi, even though they might be a bit strange
and awkward in the beginning, encourages languages to develop in the
area of communication technology.

When new technologies emerge, technology-specific language finds
its expressions at the same time. Some terms translate into native corre-
spondences easily; some remain in English. Sometimes the localization
needs to invent new expressions. For example, General Packet Radio Service (GPRS) will be localized by Hindi script having the speech sound
similar to the acronym pronounced in English. While I was writing this
chapter, my colleagues began work on the next version of the Hindi UI.
The language will be modernized this time around, aiming at a better
combination of understandability and respect for the appropriate Hindi
expressions.

Messenger

We used contextual inquiry and other qualitative contextual research methods in cultural end-user research and in direct support of concept creation. As the example from India shows, the results from these studies can be very versatile. They illuminate

- Practical requirements that arise from environmental circumstances
- The cultural values relative to specific product categories
- The products' socially and economically shaped roles in consumption
- The concepts of privacy and trust in different group structures
- The rules of socializing, such as allocation of responsibilities in a discussion
- The perception of time and space
- The idiosyncrasies of language

The results have been felt, heard, smelled, and seen. The places have been visited and the people met. This firsthand experience creates excellent preconditions for design. Context knowledge directly suggests product improvements, and provides an easy way of assessing the relevance of ideas from other sources. The bottleneck in contextual research arises from failure to communicate the results to people who haven't been involved in the actual research. The research team must ensure that the right people in product creation get the right results. As the scope of findings is very wide, the ones who are able to utilize them are spread around the organization. Further, designing based on the data requires very good and very accessible documentation and presentations. But standard written research reports can hardly mediate the experiences, subtle impressions, and emotional commitment to the end users' desires and demands that the research team has faced personally. Text can be enlivened with quotations, images, videos, charts, and other means, but that path gives out fairly quickly. Common workshops with designers and other stakeholders who need the information are a more fruitful way of communicating the data and the design imperatives implied by these data.

The author organizing observation results.

From the researchers' point of view, this means that the project doesn't wrap up on the day the report is submitted to the company database. The researcher stays for a while as a living representative of the people and the culture where she or he has been. The researcher in effect becomes the messenger.

References

1. R. Westrum, *Technologies & Society: The Shaping of People and Things*, Belmont, Calif.: Wadsworth Publishing Company, 1991.

2. K. Konkka and A. Koppinen, Nokia Mobile Phones, Finland, "Mobile Devices—Exploring Cultural Differences in Separating Professional and Personal Time," in *IWIPS Proceedings,* Baltimore, Md., 2000.

3. H. Beyer, and K. Holtzblatt, *Contextual Design: Defining Customer-Centered Systems.* San Francisco: Morgan Kaufmann, 1998.

4. S. Ruuska, "Mobile Communication Devices for International Use—Exploring Cultural Diversity through Contextual Inquiry," in *IWIPS Proceedings,* USA, 1999.

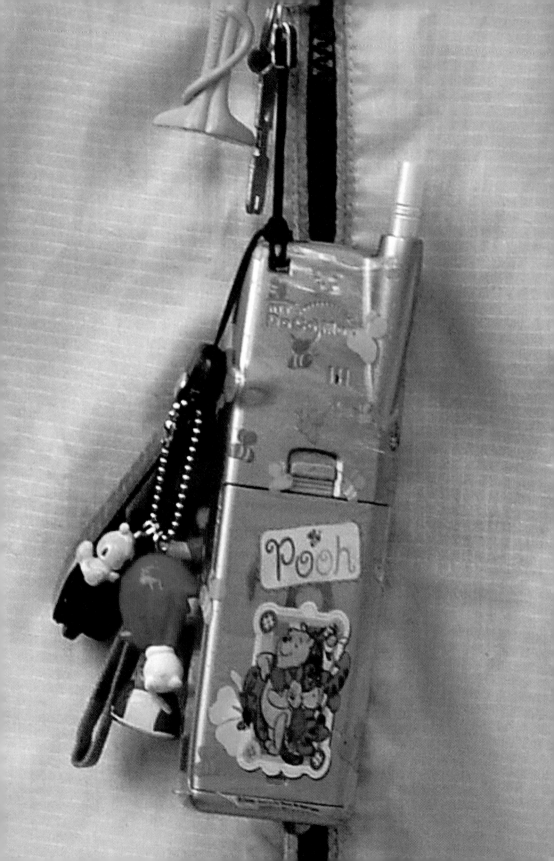

Riitta Nieminen-Sundell and Kaisa Väänänen-Vainio-Mattila

CHAPTER 5

Usability Meets Sociology for Richer Consumer Studies

Forms of communication shape our society: people communicate with each other in a mediated way via communication tools. Mobile phones, multimedia terminals, computers, and other new digital tools have entered the realm of human communication. New kinds of gadgets and services will be launched as digital convergence progresses. The user interfaces (UIs) of these products also mediate and shape human interaction. Thus they form a node for social interaction on a macro level in society. In other words, they either support and enhance or restrict the possibilities of communication.

UI designers are the people who enable or discourage expressions in technology-mediated communication. They have to be aware of the forms of interaction a UI offers. To do this, designers need to understand users holistically. The best way to approach users is with multidisciplinary methods that go beyond the ways designers have traditionally collected user intelligence. It is not enough to test products and concepts for usability; the whole development cycle should hinge on understanding the consumer. Sociology applies user research approaches that can help widen perspective on users. This chapter explains some sociological approaches that can enhance usability to meet evermore competitive product development demands.

Communication Glues Society Together

Society is elusive. Where is it? We instinctively locate it in joint activities; the way people connect with and contact others constructs the culture and society. In many societies, digital tools such as mobile terminals can mediate a remarkable proportion of human-to-human interaction. For example, in Finland, the adoption rate of mobile phones is more than 60 percent and still rising. Among young people aged 15 to 24 years, the rate is 90 percent. In the Finnish society today, there is a prevailing expectation that people have mobile phones—a social norm has been born, in other words, according to which we should be reachable most of the time.[1] Other expectations have been more subtly conditioned by the fact that many fixed services are being offered in digital and mobile forms, such as e-banking and electronically transmitted news. Television programs offer audiences a chance to participate via short messages from their mobile phones.

Digital products have assumed an important role as channels by which we reach out to other people. Thus the way in which the product mediates communication becomes crucial. How the product is used, by whom, and in what kind of situations is not purely in the hands of the users but is shaped by the designers who have preconceived its use.[2] Consequently, ease of use is key in new communications tools, to allow equal access to communications across a society irrespective of individual differences in technical literacy.

Short Message Service (SMS)

A service used in a mobile communication system by which a user can send or receive short messages—up to 160 characters—in textual form. SMS, as it is generally known, has become widely popular in Europe and the Far East since 1997, although the technology has been around since 1992. In November 2001, 26 billion text messages were sent over the global GSM network during that month only. Most SMS messages are sent person-to-person as simple text (e.g., "Meet me at the bar, 17:30"), but SMS also supports mobile information services such as news, sports, stocks, weather, horoscopes, chat, notifications, and downloadable ring tones and icons.

How the structures of the user interface force or enable individual users to communicate in a coherent way is particularly interesting. In a way, the user interface launches the terms or the *grammar* for a communication culture. For example, if SMS text messages offer only 160 characters, the communication is shaped accordingly. In advanced mobile phone markets, as in Scandinavia, Germany, or the Philippines, a whole new culture has been born around text messages as a response to the limitations 160 characters impose. Even though this restricted form of messaging may be criticized for its degraded functionality and user interface, it has obviously met its users' needs successfully.[1] The examples from the Philippines show how the limitations have actually helped create a new kind of shorthand; for instance, "CU L8R" stands for "See you later." This form of compression lets users fit more information into the available space. Restrictions set by the product designers are thus turned into a starting point for creative new expressions. Vocabulary model-based solutions (see Chap. 7 for an introduction to these) to improve text entry can speed up writing while the users stick to the standard language, but as we've seen, they don't. So, by introducing intelligent text entry algorithms in the belief that we're enhancing the messaging experience, are we actually homogenizing and restricting the users' creative expression?

It is not only exciting new forms of language that make SMS so relevant for social analysis. The meaning of SMS lies in the possibility of reaching other people wherever and whenever, and in a more direct, personal, and unobtrusive way than ever before. Users can overcome spatial and temporal limitations, which gives them more reach over their environment. Real-time communication is also more spontaneous, and the feeling of closeness is supported even though users are physically separated. It can be said that conceptions of time and space are transformed by use of the mobile phone. This more flexible relation to time and space is expressed, for example, in last-minute postponements of meetings. It is possible to plan less and improvise more.[1]

In sum, the user interface, the features, and the applications—the whole product—organizes communication and thus the society, insofar as the society resides in popular activities. It both supports and reproduces social relations that are in a sense embodied in the user interface.[2] In this way, material and virtual artifacts such as mobile phones both create and reflect social interaction. This, too, makes them worth closer scrutiny.

The new reachability has altered the nature of everyday life and society. Even though it was not obvious at the start, there has been a real need for silent, asynchronous, brief, and informal communications between persons and groups. We simply lacked the means. Responding to that latent need has created a new kind of societal bond. In this way, mobile phones with their user interfaces are a determining factor in constructing the way a mobile information society communicates.

Of course, the role of text messages has been recognized by the industry. People who prefer textual information as the main carrier for conversation can now choose terminals that enable longer messages. In the latest phone models, text messages can be sent to many recipients at once, and they accommodate more characters, which permits the use of formal language instead of short acronyms. Naturally, using a new kind of language may be more fun, and people who have enjoyed coining new terms and cryptic formulations instead of writing full sentences will probably keep doing so. It is up to users to decide which kind of discourse they prefer. The product offer has to support different kinds of users with varying tastes, although it is evident that limiting characters limits expression overall.

How, then, to research these new communication phenomena? Should it be interaction researchers or marketing researchers who delve into questions of use?

Context Shapes Communication and Users

The discipline known as *usability* or *human factors* has traditionally drawn its theories mainly from cognitive sciences. This field of research has largely assumed that individuals, namely the users, are subjects who can be studied in laboratory settings. The studies have concentrated on cognition—how individuals learn to use systems and how they perceive items in the user interface. The user is assumed to be rational and possessed of more or less clear performance goals and is also considered to be a universal creature in the sense that findings in one country or in one user segment should apply to other segments—at least when it comes to principles in perception. Even if some differences have been detected and acknowledged, the user has most often been regarded as having a rather coherent set of skills when it comes to interaction. Research has accordingly emphasized issues such as individual's perception, memory capacity, and problem-solving strategies.[3]

The discipline of human–computer interaction studies the structures of users' activities when they are trying to reach a defined goal. One such task analysis method is GOMS (goals, operators, modes, and selection rules).[4] It models user's task flow in an existing UI or prototype, and forms a basis for finding an optimal solution to the problems provoking a user's goal-oriented activities (sometimes called *procedural knowledge*). The UI design is often based on an "ideal," or the most efficient way of performing defined tasks, as determined in an expert analysis of the system. Usability tests or *cognitive walkthroughs*[5] in laboratory settings are also frequently used to gain insight into the efficiency, effectiveness, and users' subjective view of the system, usability goals defined in ISO 9241-11.[6] These approaches emphasize the instrumental nature of the interaction and represent humans as information processors.

Applications in question have expanded from desktop data processing to mobile communication and service consumption. These broader forms of action and interaction call for wider criteria in researching the usability and desirability of bundled applications, services, and products.

Although contextual and emotional aspects of usability are now being scrutinized, usability still often concentrates on the individual's performance on a device and his or her subjective report of satisfaction. These methods do not delve deeply into users' lifestyles. The new communication patterns which flourish on the street do not easily travel into laboratory settings. Sociology and cultural studies, on the other hand, try to grasp the dynamic aspects of human action. They see individuals as multifaceted *subjects* in a cultural context full of socially shaped meanings. Because we need these wider perspectives, Nokia researchers have incorporated sociological methods in our standard design practices. It is important to see users as active individuals whose actions influence, and are influenced by, culture and society. By observing and interviewing people in their real settings we can learn about communication in everyday life.

Sociology brings the social nature of both users and products into focus. In fact, sociology "problematizes" the whole concept of a user. In cultural sociology, the individual or the subject is captured at the cross-

On a slope

By a sandbox

In a bus

Confused

Frustrated

Ignorant

Graduated

Married

Retired

Stressed

Concentrating

Relaxed

Sketches drawn in 2000 for scenario manuscripts in a Nokia Research Center project that addressed future group communication services. *[By Keinonen (2000).]*

With babies

With kids

With teens

At school

At home

At work

While being served

While giving feedback

While waiting

Alone

With friends

In society

More situation sketches. *[By Keinonen (2000).]*

roads of various cultural streams, opinions, and manners of discourse. The power of multiple values, practices, and contexts in the subject's immediate environment put this individual into situations where action is not always so coherent. There are fractional and subtle meanings that do not necessarily resolve into either-or options. For example, some computer users may both like and hate the terminal they use, or may be both capable and incapable of using it. They may see the value of its functions but also consider the terminal ugly or even immoral. They may express contradictory opinions in different situations.

Even though technology and its social implications have divided opinions between—and within—individuals for ages, attitudes toward mobile communication are just starting to get interesting for observers. Chapter 12 illuminates future trends in mobile communication, which are perfectly characterized by a single word: contradictory.

Sociologists observe that people take different stances in different situations to better fit to the expectations of others. They may be polite in one situation and arrogant in another. The *context* influences which direction users go. Human action is thus open in nature—it cannot be reduced to laws of behaviour even if biology and cognition set some restrictions on the scope of action. Instead of deterministic rules or natural laws in human action, we find a range of possibilities, which users can actualize. Thus it cannot be stated that a certain UI structure always provokes a certain kind of reaction. This is not to deny that general principles in human perception and cognition, such as the principle of proximity, give some restrictive guidelines for design. They are not, however, sufficient to guide product development. In real life, much UI design addresses the appropriateness of features, including their priorities and symbolic meanings, for a given user segment. These decisions cannot rely solely on the usability findings or ergonomics. Instead, we have to examine human action in its cultural context to see how usability intertwines with other product preferences.

Thus "context of use" is one of the main extensions of cognitive user perception in modern usability as defined by Bevan and Macleod,[7] "Usability is a property of the overall system: it is the quality of use in a context." The concept of context includes the physical, situational, and, to some extent, social implications of a product design. *Physical con-*

texts naturally influence, say, the way we can interact with the physical terminal—noise level or weather limits the mobile phone user's ability to use the phone successfully. *Situational contexts* are functions of several factors:

- What kind of culturally valid opinion sets are available to the user (are mobile phones accepted in this situation)?
- What kind of expectations must the user meet (should he call or send an SMS)?
- With whom is the user going to communicate (would the person being contacted prefer formal or informal language)?

Let's look at some opinions available for technology users. These may include "technology optimism," which promotes or requires that the user assume a skilful identity. Users who feel adept emphasize their skills and knowledge, even if they actually find some things difficult. Correspondingly, a user with technology resistance may claim to reject all gadgets or (deliberately) lack all skills related to a certain digital tool, thereby emphasizing and attesting to the critical attitude. Attitude may have an effect on behavioral performance measurements. It is key to see users as individuals with attitudes, interests, and activities to understand how their interactions with products are part of their everyday routines. According to these attitudes and preferences, users can be segmented into groups with more or less coherent behaviour.

Different Users Want Different Things

To understand the various cultural positions users take, researchers separate technology users groups according to their attitudes toward technology (or other product-related characteristics) to clarify differences within the market. For example, a very common practice in business contexts is to distinguish between techies or innovators, early adopters, early and late majority, and laggards.[8] This kind of segmentation emphasizes differences of attitude and skill, but does not study users in contexts of use or as members of cultures. Thus a segmentation based on technology adoption only lacks a means of observing the changing nature of the innovation in question over time.

In the mobile phone business it is crucial that we understand not only the cognitive needs of individuals but also the broader issues arising from the social contexts in which our products are used. In the next section we will explore innately incoherent individuals and fragmented user populations from the viewpoint of sociological research tradition. Research in this tradition aims to enhance usability by problematizing the user and to expand it with contextually aware practices.

Products Change after the Launch

It is not only consumers we must understand in their social settings; it is also the products. Products are created and consumed within a culture. Manufacturers, advertisers, and users invest them with different meanings. Thus the product has a social dimension conferred on it in the form of possibilities of use; only some kinds of interactions are possible. The social dimension also includes meanings attached to product users that shape each user's identity through possession and use of the product.[2,9]

Sometimes the meanings given to products can be so distinctive that they multiply the product into different *cultural artifacts*. When bicycles were introduced, for example, they were both strongly opposed and passionately loved by the consumers of the late nineteenth century. The antagonists saw the vehicle as a dangerous and grandiose gadget meant for showing off, and the protagonists saw it as an exciting, sporty appurtenance for adventurous men. Products can provoke controversy and user ambivalence—they do it all the time. In the case of the bicycle, cultural debate about the meaning of the thing abated and the product got its more-or-less stable meaning as a means of transportation (although today we can see the return of a racier image in form of the mountain bike).[10]

What's radical here is the idea that products *keep changing after the launch phase* (Fig. 5.1).[11] In the case of mobile phones, it means that the phone has travelled a trajectory from an expensive showoff tool to an everyman's communication platform. The meaning of the phone (and its relation to the consumer) is in a way renegotiated in the daily context

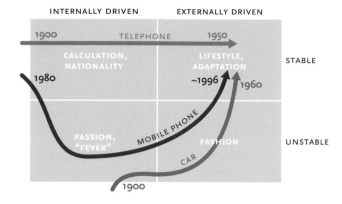

Figure 5.1 Products keep changing after the original launch phase. *[From Pantzar[10]]*

again and again. It is also evident that every new-product launch—new models within the same product category—challenges the role the phone plays.

The same people may have different ideas about the very same product at different points in time, partly because the product tends to become more acceptable over time. When the short message service (SMS) was launched, initially nothing noteworthy happened. Yet when SMS began to be widely used, GSM phones exhibited a new usage pattern. The phone became more than a phone—it was now a textual communication tool that allowed people to keep in touch in a way never seen before. A whole culture was created by users for users. The ways people relate to each other changed, and users learned usage behaviors that they could not have imagined before.[1]

When SMS was new, it was unfamiliar to many users. Elderly consumers may have seen it as especially strange. Now we see more and more senior consumers using text messages, an application first conceived for teenagers. In this way, users may slowly adjust to the product and thus relate to it more positively with time. With new users come new meanings for the product, and meanings in turn change attitudes, which definitely influence consumer perceptions of the product.[11] Manufacturers can't afford to ignore new meanings created in the marketplace when conceiving new-product versions. If they do, they may expect to be surprised by consumers rejecting the new product.

To recap, the social nature of users—their heterogeneity, for example—is not the only new subject of study for usability investigations. The products themselves have a less coherent nature with a changing image. Culture as a source of meanings should thus be included in the analysis and acceptance of products. But, instead of just adding "culture" to the traditional model of human–computer interaction, a more radical step is to reconceive the whole product development chain as a locus for creating and reproducing culture. Socially shaped meanings are inscribed in or attached to products, which then are transferred to the everyday life of the users to become what they will.[2,9]

Sociology and Usability Research

The social aspects of interactions, users, and products invite designers to learn more about the communication culture in particular, and consumption in general. To mobile phone designers, the challenge is huge—no phone has only one kind of user in only one kind of context. There are always both experts and novices working the phone in both noisy and quiet settings, to which they bring various cultures with varying opinion sets, values, and learning histories. *All* of them bring their life, social circle, work demands, and aesthetic taste to the interaction. In this sense, the product never once reaches a stereotypical user with a plain cognitive mind. UIs thus have to serve a huge number of individuals, all different.

That said, contexts do not preclude patterns of use, or even valid usability test results. An interaction with a product is based on the user interface that structures communication. Design solutions are thus most crucial because they impose similarities on the communication, no matter which context they end up in.[2] Moreover, interaction has both context-free and context-sensitive aspects.[12] It is the context-free structure of the user interface that defines possibilities for the context-sensitive aspects. Context-free aspects, of course, can still be tested in laboratory settings with a reasonable expectation of generalizable results. Context-sensitive parts of the interaction can be evaluated only "on location." The key is to combine various methods for a fuller picture of how a product fares among its target audience.

With respect to practical research activities for product development, the arguments above have two kinds of consequences. The first affects the quality and amount of data gathered in studies. Usability tests that treat the user as a universal mind do not extrapolate findings into a larger context. It is important to widen the settings of a usability test by including, for example, a section of questions to define the user's market segment and that individual's position with respect to technology in general. It is also important to get beyond the obvious statistical data on demographics. The ultimate goal is to collect information about a user's socially shaped attitudes toward the technology or product category in question.

The second consequence affects testing locations. Usability studies have traditionally been conducted in laboratories, in a controlled environment, like an experiment in natural science. This kind of setting does not reflect the myriad of changes inherent in the usage situation or in the user. Bringing the test into a real-life context has been one further step taken by many practitioners since the late 1990s.

All in all, sociology widens the user interaction study by introducing three major themes:

1. *Change.* The individual has different "development phases" as a consumer. His or her preferences change in relation to the socially shaped meanings that products acquire.
2. *Culture.* Culture is a set of practices that shape our interaction with machines and thus influence their usability.
3. *Fragmentation.* There is variety in human action that cannot be explained by reference to cultural practices alone. People differ in many ways, and we cannot test them all. Therefore the way users are segmented for studies is crucial to what we learn.

In mobile communications, we can never hope to define all attributes of an environment, nor all situations in which the user will communicate or carry out other mobile tasks. A research mix has to be created that can uncover even the surprising aspects of the new communication phenomena. These phenomena place requirements on the user interfaces as well as on the product's feature set. Researching them solely through predefined tests with given tasks does not help us learn what to learn about the users.

Research Methods for a Broad Understanding of End-User Needs

From the discussion above we can now draw some conclusions about how usability research might be extended to cover the fragmented nature of users and user groups:

1. *Longitudinal acceptability analysis*—how domestication shifts an innovative product into the category of everyday consumer durables. We also need to investigate the user learning curve for a product or a prototype, as well as the "boredom curve," or how soon the user will grasp the (intended or unintended) challenges of an explorative product.

2. *Contextual user needs studies*—identifying material differences within the user population and uncovering incoherencies in action in arising from different usage contexts. Furthermore, contextual research is needed to understand the full range of situations in which users must be able to use the product in a satisfying or even *enlightening* way. Ethnography offers useful methods for investigating the in-depth needs of users in contexts, and for designing products to suit those contexts. The contextual inquiry method by Hugh Bayer and Karen Holtzblatt is a well documented method for these purposes.[13]

3. *Cultural research or ethnography*—utilizing in-depth interviews to construct an account of possible user perspectives instead of forced either-or options. Culture does not mean just users' language or the formal rules imposed on them by virtue of nationality, but also the values, habits, and thinking structures that organize their everyday activities.

All studies should be conducted with well-defined user segments. Researchers should not settle for investigating users and potential users solely by means of lab tests or questionnaires, which prompt them to answer in predefined ways. People may have many "hidden" needs of which even they are not quite aware. In-depth interviews and contextual observations are thus routes to broad, qualitative understanding of end users and their overt and latent needs. However, if you do create a research method mix, its validity has to be evaluated regularly to make sure that new phenomena are being caught by its components. Change in the marketplace is quick, so designers have to be alert to pick up any relevant changes that new products and related communication practices bring in their wake. The constant interplay of the product and the user's culture shapes both of them, and should be a standard part of defining user requirements.

Better Research, Better Products

Although mobile phones and other new digital communication tools can be seen to *mediate* human communication, they also *shape* it. As argued here, user interfaces have a rich relationship to the social activities, and they themselves are an element in the socially shaped and socially shaping interaction. *UIs per se represent society for us.* Thus, *we actually are carrying a piece of society in our pockets* each time we walk out with our phones.[1] And UI specifiers, product concept developers, and software engineers—they all design society too. This is why UI designers have such an important role in defining the mobile information society.

Naturally, the relationship between the UI and the user (or the whole culture) is dynamic. The modeling of existing communication needs necessarily intertwines with innovations that reshape communication. Innovative products always offer something more than is there today. Innovations change communication patterns, which for their part then change product requirements. In this sense, communication patterns serve as a starting point for innovations, and they should never be underestimated.

Product development processes and associated end-user insights must be tempered by an awareness of changes in the human–machine interaction and communication culture. The methods mentioned in this chapter—longitudinal, contextual, and cultural end-user research—offer a basis for practical development work. They also bring usability practice closer to other consumer-related studies. Forming a richer picture of the people using our products is a core challenge. We will never be finished painting this picture. And because consumers change, products have to change as well. There is a reason to look very closely at human interaction and what it demands from the user interface. The pace of the digital marketplace generates new kinds of digital product launch challenges for consumer researchers and UI designers, and forces them to understand emerging cultural patterns to offer better products in the next round. What was great once soon becomes outdated. Today's SMS will not be enough tomorrow.

References

1. T. Kopomaa, *The City in Your Pocket: Birth of the Mobile Information Society.* Helsinki: Gaudeamus, 2000.

2. S. Woolgar, "Technologies as Cultural Artefacts," in *Information and Communication Technologies. Visions and Realities,* Dutton, ed. Oxford Univ. Press, 1996.

3. M. Gardiner and B. Christie, *Applying Cognitive Psychology to User Interface Design.* New York: Wiley, 1987.

4. S. Card, T. Moran, and A. Newell, "The Keystroke-Level Model for User Performance Time with Interactive Systems," *Commun. ACM* **23**(7): 398–410 (1980).

5. C. Wharton, J. Rieman, C. Lewis, and P. Polson, "*The Cognitive Walkthrough: A Practitioner's Guide,*" in *Usability Inspection Methods,* J. Nielsen and R. L. Mack, eds. New York: Wiley, 1994.

6. ISO 9241-11, *Ergonomic Requirements for Office Work with Visual Display Terminals (VDTs),* Part 11: *Guidance on Usability,* 1998.

7. N. Bevan and M. Macleod, "Usability Measurement in Context," *Behavior and Information Technology* **13**(1, 2): 132–145 (1994).

8. E. M. Rogers, *Diffusion of Innovations.* New York: The Free Press, 1983.

9. G. McCracken, *Culture and Consumption. New Approaches to the Symbolic Character of Consumer Goods and Activities.* Indiana Univ. Press, 1990.

10. W. E. Bijker, *Of Bicycles, Bakelites, and Bulbs. Toward a Theory of Sociotechnical Change.* Cambridge, Mass.: The MIT Press, 1999.

11. M. Pantzar, *Kuinka teknologia kesytetään. Kulutuksen tieteestä kulutuksen taiteeseen* (How to Tame Technology). Helsinki: Tammi, 1996.

12. H. Sacks, E. A. Schegloff, and G. Jefferson, "A Simplest Systematics for the Organization of Turn-Taking for Conversation," *Language* **50**(4) (1974).

13. H. Beyer and K. Holtzblatt, *Contextual Design: Defining Customer-Centered Systems.* San Francisco: Morgan Kaufmann, 1998.

14. K. Kuutti, "Hunting for the Lost User," keynote speech in the Cultural Usability Seminar, 24th April 24, 2001, paper 2001-04-25; *http://mlab. uiah.fi/culturalusability/papers/Kuutti_paper.html.*

PART 3

Design and Research Intertwined

The stories you'll read in this part of the book are real-world examples of mobile UI design challenges, such as squeezing text input tools into minuscule spaces, localizing user interfaces (UIs) hand-in-glove with UI-style development, devising a way to research complex human–machine interactions for users who make phone calls and drive cars at the same time, and finding out what customers dislike most about using mobile services. The projects and the problem solving that project teams actually do will also give you a new appreciation of how many corporate functions contribute to usability engineering.

Chapters 6 to 11 reveal the range of objectives, approaches, and points of view espoused by usability and interaction design professionals at Nokia. Usability as a corporate function includes focused and comprehensive problem solving, creative design, hard decision making, leaps of faith, and thoughtful management to facilitate the work. It takes a judicious selection from the approaches listed below—and sometimes all of them—to make a mobile project cohere.

> *Engineered.* Theory is both the starting point for new hypotheses and the framework where novel concepts can be accommodated and pursued. It is the momentum behind cumulative progress in science. Interaction design and usability engineering projects place a lot of weight on designing and engineering solutions and validating those solutions with user tests, but analyzing the phenomena behind the results gets less attention. Industrial usability research does not focus on turning data into theories, at least not on the basis of our experience. Getting something to work is our primary method of learning why it actually functions—or why it doesn't. An obvious exception to this rule occurs when the lessons learned can be immediately iterated

and turned into functional solutions to be applied in the next proto-type version or in the next product. Usability operates primarily in a constructive engineering mode. Why is that so? Isn't it self-evident that understanding the underlying principles of human–machine interaction would enhance our chances of success in the following rounds of design? In UI research oriented to practical product devel-opment, we encounter numerous variables related to human cogni-tion, motivation, usage context, technology, and so on that may influence user experience. They can—in the best case—be identified, but to define each variable's role in a specific problem is beyond the scope of present understanding. Miika Silfverberg's report (Chap. 7) is a good example of the dilemma. He presents well-justified assump-tions in the beginning of the project, but even so is surprised by what turns out to be the biggest issues as revealed in users' actual behavior when interacting with a prototype.

Intuitive. Usability organizations must ensure that usability and UI design knowledge is captured and catalogued, but the case-dependent nature of the discipline makes that difficult. Without cumulative theories—or professional conventions and rules, to be more modest—the people involved are the best conduits of knowledge in the lab. Their experi-ences with successful and failed projects turn into a kind of tacit knowl-edge or developed intuition. They become experts. They will gradually be able to contribute more focused studies, more rewarding new initia-tives, or more helpful UI solutions. Even though it may seem somewhat contradictory to the principles of user-centered design and multidiscipli-nary cooperation, the knowledge of the most experienced designers and usability experts is one of the most profound strengths of a corpo-rate usability organization. When it's impossible to test every detail, the opinion of the expert counts for more. The experts, in return, need opportunities to nurture their expertise.

Midpriced. Advocates of "discount" usability have argued their case con-vincingly by proving the validity of its results—four test subjects are enough; usability tests with lo-fi (low-fidelity) prototypes will tell you what you need to know. We accept their findings, but we also recognize that discount isn't always the best value for the money. There are two main routes to improving the validity of usability research—or to slip into the midprice or even premium range: (1) improve the fidelity of the stimulus material (i.e., simulations and prototypes) and (2) construct more valid test settings with reference to sampling and context. Both will quickly outspend a true discount budget. Chapters 6 to 11 showcase a couple of cases where discount methods were rejected even though the projects were in their early phases: Pekka Ketola's contribution (Chap. 8) about internationalizing and localizing a UI style illustrates one reason; Turkka Keinonen's contribution (Chap. 9) about in-car user interfaces shows another.

International. Global perspective on usability is a recurring topic in this book, and it is also one of the main challenges in making reasonable tradeoffs between comprehensive engineering approaches and discount usability. Simple methods can grow into remarkably large-scale efforts as the evaluations are carried out in several countries on separate continents. Translating the stimulus material, organizing travel arrangements, and teaching local moderators—it all takes time. The discount usability mindset starts to turn into something else when your goal is to test the understandability of terms and navigation sequences in several language areas. Researchers must heed the relationship between the cost of their efforts and the usefulness of the results, as Pekka Ketola does. The tasks that require world tours must be distinguished from those for which testing with some colleagues in a neighboring department suffices.

Real-time. Time is a crucial factor in UI design. It is even more critical when designing user interfaces to be manipulated while driving. Drivers' attention-sharing behavior quickly becomes the main usability challenge. A user interface for a car system itself must adhere to severe limits on how much of the driver's visual attention it requires. Thus, designing for *attention sharing* calls for trials and test tools that deliver inputs at the pace of real products, and that provide realistic tactile and auditory responses. Although driving is an extreme case underlining the importance of a users' attention-sharing performance, the issue is common in mobile UI design. Mobile UIs must be robust in the face of continuous interruptions and usable in recurring dual-task situations. Consequently, your usability evaluation should be able to assess the robustness of proposed solutions, which puts tough requirements on the prototypes.

Behavioral. Contextualizing usability approaches in user studies was emphasized in Part 2. This part (Chaps. 6 to 10) deals with design and concept creation, and here we often have to accept less context and more isolation of the human–machine interaction in a research setting. Behavioral usability research can focus on a specific aspect of human performance: perhaps the ability to learn a new input technique or to divide attention between primary and secondary tasks. When this is the case, the projects may rely on laboratory measurements and company-internal subjects—or at least such reliance can be seen as an acceptable tradeoff for efficiency. Typically the aim in these test settings is to compare several new design solutions, or a new solution to an existing one, with quantitative criteria. The tradeoff pays the researcher back by providing quantitative results relatively soon.

Contextual. Another category of design and research approaches aims at integrating solutions for a coherent user experience. Screen designs and interaction conventions must be combined to form a unified concept for a mobile phone UI, and various technologies and services must be conjoined to deliver mobile Internet services, as Pekkarinen and Salo demonstrate (Chap. 10). An acceptable or interesting solution is judged not only by reference to an existing product or an alternative concept but also, above all, by the users' expectations for utility and pleasure. Test settings accordingly have to be more contextual than the isolated behavioral measurements applied to the study of specific aspects of interaction. Assessments are carried out in the field with external subjects and with open-ended research approaches.

Responsive. Usability activities in industrial settings are typically related to the product creation process. The activities may provide support to product programs or may actually be a part of design-related decision

making. In these cases usability functions in a *responsive* mode, react-
ing to the challenges posed by emerging technologies and the design
of new products. Product programs set the schedules and determine
the necessary level of detail for the results. It's essential to come up
with concrete proposals or definitive answers as to the superiority of
one of the alternatives. In Chap. 11 John Rieman tells us about one
way of doing responsive usability engineering.

Proactive. The remaining examples concern concept creation exercises or
interaction research preparatory to making implementation decisions.
UI designers and usability specialists may meet the project objectives
and approaches relatively independently, as they see appropriate.
They are given an opportunity to take the initiative and propose solutions on the basis of their understanding of optimal user experience.
Usability research of this sort functions in a proactive mode.

Managed. Constructive, midprice, behavioral, and proactive mode—the
whole spectrum of usability approaches is needed to balance the
speed, costs, and reliability requirements of usability engineering. Corporate usability groups must have enough flexibility in its toolkit of
methods to cover the range. Management must understand the business, the user interfaces, and the importance of usability to make the
right choices among possible modes of operation. The corporate
usability function will not be organized effectively and function
smoothly unless it has facilitators and leaders. User interface management ensures that expertise in a usability organization is utilized to its
full potential. To realize this vision, the UI design manager must have
the skills to manage the team and the project, quality, documentation, and the product itself, as Christian Lindholm and Turkka
Keinonen show in Chap. 6.

Short
text
messaging
Select

Text
Messaging
Select   Back

Messages

Select   Back

1.Messages
2.Register
3.Profiles
Select   Back

Select

Back

Christian Lindholm and Turkka Keinonen

CHAPTER 6

Managing the Design of User Interfaces

Nokia has been strategically managing the design of mobile phone user interfaces since the early 1990s. Design methods, tools, and training approaches have been developed continuously, and numerous gadgets and simulations have been evaluated in our labs with varying success.

The UI management teams at Nokia have learned to operate in an extremely dynamic and independent environment with sometimes conflicting expectations. The rapid technological development in mobile communications has influenced mobile device usability a lot. To get a sense of what this means, one needs only to look at the advances in LCD technology as it has transitioned from character displays to dot-matrix displays and now into active-matrix color displays, all in less than 7 years (at the time of writing).

Since mobile products have to be small, it is often not possible to reuse ideas directly from the desktop UI world. *Small user interfaces do not scale* seems to be one of the natural laws governing our work. The mobile is a totally different medium from the desktop PC. Among our customers is a new, very capable generation of young users who are totally at ease watching MTV and playing on their PlayStation at the same time. Satisfying both these "screenagers" and the demands of the maturing population is a real challenge.

Since the mid 1990s, Nokia's strategies have been driven by three different organizations: product marketing, concept creation, and R&D (research and development). We have searched for the "right" model of organizing UI design activities, and we have found that each solution has inherent benefits and deficiencies.

Under product marketing, UI strategy development aimed at setting the vision, strategy, and roadmap for designing user interfaces. Documenting these results into requirements and communicating them to R&D were its daily tasks. The allocation of responsibilities between product marketing and R&D resembled the familiar relationship between customer and supplier, which eliminated some of the communication problems existing between different experts. Another benefit of having product marketing take the lead in strategic UI development was that the top management could directly influence the UI strategy.

Concept creation's approach to UI management was to stimulate and sharpen the vision, and then build a strategy to implement it. Nokia decided to set up a small Product Concept Group responsible for advancing the product concept, forming a kind of "think tank" or, more evocatively in Finnish, an "idea furnace." Most of the experts in this group came from the research side. Its first challenge was that the gap between reality and vision became too wide—the urgent reality machine of schedules and resource allocations was missing. The second challenge was territorialism: people started to think that the group automatically had a monopoly on good ideas and all the others were inferior.

For its part, Nokia R&D set up a global cross-functional group of user interface experts called the User Interaction Group (UIG). It enlisted all user interface designers throughout the corporation. There were six different disciplines in the UIG specification:

- Interaction design
- Localization
- Graphic design
- User interface platform (the team responsible for strategy, roadmapping, and requirement gathering)
- Sound design
- Usability

The benefit of this cross-disciplinary organization was that it united the UI designers from different locations and backgrounds to create easy-to-use products, and communication within the global group was excellent. Problems arose, however, in communicating with implementation and product marketing. The UIG was seen as the third leg in product creation, and three was too many. The UIG organization was too far from product planning and, even though technically in R&D, still not aligned with implementation. The mandate of the UIG was to ensure that an interface was easy to use, not that it was easy to make. In the end, the UIG served as a "university of UI design" during Nokia's intensive growth phase. The fact that the user interface people sat down together allowed them to learn quickly from each other.

Through these different approaches to managing UI design we have learned by doing. Our experience has given us a pack of tools to be applied in managing UI development in an innovative expert organization. Some of the most essential tools are introduced in the following paragraphs.

Vision Management

The UI creation process starts with a vision, the discovery of undetected user needs, a foreseeable change in the industry, or simply a spark of inspiration. The initiative can come from anyone. UI management is responsible for recognizing fruitful initiatives and arranging for ways to nurture them. Getting started doesn't depend on piles of reports; a few presentation slides with sketches or text notes, or even a forceful sentence or two, may be good enough.

What is paramount is that the message is presented *in person* to the *right audience.* The challenge is to find the people who see the value of the idea, who have a demand for that kind of innovation, who have the power to drive the idea, and who have the expertise to turn the ghost of an idea into a tangible concept. Personal presentation of the initial ideas enables instant adaptation and modification of the vision in order to increase the stakeholders' commitment and interest.

Vision-driven design is not based on comprehensive analysis of background information. It aims at recognizing an attractive goal for development, and then the goal guides the project's progress, putting details into their appropriate places and giving them the weight they deserve. The vision, even though its origin cannot be defined in a completely satisfactory manner, is typically based partly on professional understanding and ideals, and partly on the task-related criteria. It is possible and necessary to continue gathering more and more task-related information during the process, but if the original direction is not the right one, no amount of effort will help.

In introducing a vision, it is not important to paint a complete picture. A vision has to leave ample room for imagination. If there is nothing to spark the imagination, how can one get the audience excited? The vision has to provide the *spine*, the *goal* or *beacon* to guide the design team throughout the development process. The vision can and should evolve as more data and knowledge is discovered. But if the vision collapses then the project should be terminated. There is no point in continuing without a target, without something to give the work its gratification.

Metaphors are good tools to use in visioning. A good metaphor clarifies a vision with just a few words. The beauty of a metaphor is that it is often easy to remember, and it communicates across professional and even cultural borders. It should be rugged, yet adaptive to the situation and the audience. No metaphor is bulletproof, but then again it doesn't have to be.

Product Substitutes

Think of a car, shoes, a bicycle, and a scooter. All of these serve the same basic need to transport humans from one place to another; hence these products are all in the human transportation business. Nokia is in the communication business, and our objective is thus to create communication products; and these will be as different as a car is from a bicycle and a bicycle from a shoe, and so on. These products will become both product substitutes, as cars are to motorcycles, and product complements, as shoes are to scooters.

We do not know if a phone is a car or a motorcycle or just a pair of communication shoes. By the time we find out, we already have devices profiled to communicate some or all of these.

Product segmentation means creating focused products for specific customer segments. The individual product usually cannot be good at all things, and it is even acceptable for a product to be bad at something. In fact, the existence of a weakness can reinforce the observer's recognition of strengths. One of Nokia's senior UI design experts once used a motorcycle metaphor. He said that if you buy a motorcycle, it starts raining when you take it for a spin, and you get wet, is it a bad motorcycle? Of course not. If rain protection had been a key concern, maybe motorcycles would never have been created.

A more concrete way to describe a vision is to create a scenario. A scenario of an everyday situation encourages the audience to relate their own insights to the vision. The scenario is a story about the use of a product. It describes the people using a product, the environment in which they use it, their procedures for using it, the product's features, and the benefits it provides to users. Scenarios used in presenting visions must be elaborated to the point where a usage context can be linked with the idea. Sometimes this is very obvious from the very beginning, but in other cases scenario building may require additional efforts. Concerning the presentation of the product itself—its physical appearance and the user interface—scenarios are flexible. The product can be clearly illustrated or hidden, and we can express what needs to be expressed through the created context.

Construction plans of information society are often presented as scenarios—they can present very imaginative worlds of technology visions. It is relatively easy to create stories where the technology is perfect and all limitations and problems related to the acceptability and usability of new solutions are omitted. What makes scenario building interesting, though, is incomplete technology and a commitment to the concrete requirements set by human behavior. There's no fear of getting lost in utopia if you stay with stories where people *as they are around us right now* consume novel technologies.

The best scenarios, like the best metaphors, are simple, so that they can be remembered and passed on to a third party. A scenario has to progress in a straightforward manner. Information is given out as small, logically progressing steps. This is necessary to make services that are often complicated, easy and understandable to follow. A single scenario can combine in a flexible manner what is already known by research or past experience, and what is the innovative created content. The credibility of new ideas is often tried for the first time when writing scripts for a scenario. For some product ideas, it is extremely difficult to write a scenario; this recalcitrance may indicate problems with the ideas themselves. A technically interesting idea may face problems in finding its place in a realistic context of use.

In addition to metaphors and scenarios, a third conceptual tool for vision-driven user interface creation is a *design driver*. A *design driver* may be defined as a design objective that (1) has a very high priority in concept creation; (2) characterizes the concept in a way that underlines its distinctive properties; (3) is comprehensive by nature, affecting several aspects of the design; and (4) can be presented with one simple, clear sentence or phrase.

A good example of a design driver is *one-hand use.* The driver is easy to understand and communicate with. The decision to adhere to single-hand use has a profound effect on several aspects of the user interface design, such as the terminal's form factor, key layout, key press sequences, and the implementation of shortcuts.

Design drivers are more explicit definitions of the design objectives than metaphors or scenarios. They closely resemble design requirements but, as opposed to requirements, do not ascribe certain values to individual objectives. Instead they identify the most important dimensions that need to be optimized in a concept. A design must be driven as far as possible with reference to its drivers. Other design objectives of lesser importance will later be matched to the overall concept that the drivers have introduced.

To summarize, vision-driven user interface design is a way to direct development by highlighting desirable goals for stakeholder commitment and by giving a "soul" to the design from the very beginning. Vision guides the process and gives meaning to the details. Vision can be explicated and managed by metaphors, scenarios, and design drivers. However, visions are never fixed, but always open enough to nurture the design team's imagination.

Team Management

The Nokia UI design culture and way of working could be described as an *individualistic network-based expert organization.* This means that it is commonplace to assign the most suitable person to the task rather than take the task to the most suitable organization. This works because we constantly create and destroy the organization, leaving only the people and processes in which competence resides. And as processes are merely work methods, that leaves us with people.

Assembling the right team for a project is always a challenge. The bigger the team, the more likely it will avoid mistakes related to missing skills and perspectives, but size comes at the expense of speed. Hence the design manager usually has to balance reliability versus efficiency. Three-person teams are fantastic for fluent cooperation, and a really experienced team of three experts can create a whole product concept. Teams of five or fewer people still function well, but in our experience the adage that five people are a team and seven are a committee seems to hold true. Committees are no good for design purposes.

Favoring speed in design team assembly greatly stresses the individual designers and their skills. A three-person user interface creation team needs an interaction designer, a graphic designer, and a marketing representative. The interaction designer is usually responsible for conceiving the dialogue and for the necessary technical knowledge and simulation; the graphic designer turns the dialogue into screen designs, and the marketer identifies the customers, including their needs and their market segments. In this team there are no redundancies, but it is short of two important competencies: software implementation and localization. Experts in these disciplines must therefore be consulted on an ad hoc basis during the design process, or at least at agreed milestones. If the objective is to design a total product concept with a novel physical form and appearance, an industrial designer should join the team.

Team spirit has a major impact on the outcome. When team dynamics are harmonious and the team members know each other, then a bonding of mutual respect appears, assumptions coalesce, and ideas get reviewed and evolve at the speed of thought. To maintain the right creative and productive spirit, all members have to be contributing, and every team member needs to count on others' contribution. No free rides—only active contributors may hang around. Team spirit is enhanced by the perceived importance of the work the team does. Designers like their work to be seen. At intervals, appropriate responses to the team's achievements are important to the design team's confidence and momentum. This can be ensured by regular reviews and presentations to management.

Nokia's user interface style creation projects, particularly the ones for the high-end concepts like the Series 60 smart phone style and the Series 80 communicator UI, have been extensive efforts right from initial concept creation through prototyping, localization, and evaluation with world tours, all the way through to software implementation. Working for 4 to 5 years on the same project is a long time to maintain vision, interest, and focus, but that is the time it takes to do a completely new high-end UI style. During long-running projects it is often the case that the design team changes design midway—even several times, and this is likely to affect the consistency of the design. Perhaps the reason for allowing the change has something to do with designers' personal characteristics. The nature of development changes as projects proceed, and some people are better at kicking new things off than completing them. For these people the boredom after the major decisions have been made might be a reason to lose their motivation. Luckily there are others who enjoy striving for details. According to our experience, the takeover may not occur without hiccups. The most important thing is to ensure that the whole concept team does not withdraw at the same time, but that some members continue as consultants to the implementation part of the project. What also helps in these transitions is good documentation. In the case of Series 60 we had a very advanced PC-based simulation. At the time of launching the first Series 60 phone, anyone looking at the simulation and the actual software could tell that they were strikingly similar. Another source of help is a comprehensive UI style guide.

Nokia has recruited with very open minds. In the Nokia human factors community we have sociologists, engineers, audio experts, linguists, anthropologists, graphic designers, mathematicians, and economists. What they share is that all are working within the bounds of their own fields on the same project, and all are interested in exploring new ground. However, this breadth of competencies may create communication problems. To overcome these, Nokia has established procedures and working practices, and created mobile UI jargon that is used internally. These processes are the backbone of our work and therefore have to be constantly tuned and improved.

The concept creation competition is one such process. Vision-driven concept creation leaves much of the project outcome to the skills and creativity of the design teams. Whenever possible, it is advisable to create several competing concepts. Following a number of promising paths for a while will increase the likelihood of finding the right design. Instead of asking one team to propose several solutions, it is better to run a number of teams in parallel—typically two to six—cracking the same task. The friendly contest creates some tension to sharpen the mind, and produces better innovations.

Besides creating design alternatives, a design competition may help identify the most suitable team to continue the project. Competing concepts are realized within very tight schedules. The time constraint is important to ensure that no one involved becomes so attached to one design that he or she can't jump on board and work toward another that happens to be selected for further development. The concepts can be changed, but not the individual designers.

In a competition review the teams receive direct feedback from user interface design management, and each team is encouraged to express their opinions for and against other concepts. In this phase it is important to focus on core issues such as number of keys, type of keys, size of display, navigation model, and the layouts of core applications. Much of the detailed interaction can be engineered later. Sometimes external agencies have participated in Nokia's UI competitions. The biggest disappointments in those cases have occurred when the ratio between "icing" and "cake" has been reversed. It seems that for novice designers it is easier to do good-looking than well-working designs. The longer the experience of the designer, in fact, the more plain and stripped-down the concepts become. As long as the windowing calculations are correct, UI graphics can be added later.

The hard part of UI concept competitions is judging. Actually a winner is seldom declared, and no prize is given—unless you consider it a prize to be able to continue work on the concept toward a real product rather than letting it end up as just another layer of technosediment in the cabinet. Such competitions rarely result in one uniquely superior concept, but typically all concepts have some excellent characteristics and some unacceptable solutions. The reason for this is the lack of time needed to polish the concepts.

Conducting a design competition will add time to the process of conceptualizing a product. As a rule of thumb, two concepts take 50 percent more time and three take 100 percent more time than working with a single one. The increments come from consolidation. Usually, one concept is selected as the basis of the project and then the designers identify elements from the other concepts that will be incorporated into it. From then on the project becomes straightforward design work.

Quality Management

Concepts must be continuously tested throughout the development process. Showing concepts to human-computer interaction experts allows key problems to be pinpointed in the early phases of design. Later in the process expert evaluation can be used to complement user test for aspects that are hard to evaluate with users, such as checking usability in continuous use and assessing the influence of UI elements that are still missing. The main advantage of expert evaluation as a method of concept testing is that it is flexible with regard to presentation materials. Experts have knowledge about the product development and design process. Consequently, it is possible to get good readings from unfinished material supported by verbal qualifications, references to products and styles the experts know, and so on. An expert analysis of a concept can be done in 2 to 3 days. Since the results depend on the competence of the evaluators, the approach requires strong trust on the part of management in the expertise available to it. Without qualified evaluators, it just does not make sense to carry out these exercises.

Whenever possible, usability testing with interactive prototypes is preferred. They provide a much richer tool for capturing feedback. Even an extremely simple interactive simulation done with a low-fidelity prototyping tool can be highly effective.

User tests in themselves don't necessarily consume much time, but careful preparations such as recruiting users and planning tasks will increase the total duration. A proper usability test can seldom be done in under 4 to 5 days. (See Chap. 8 for the resources calculation for an international test round.) User testing should obviously be conducted with test users who represent the right target segment. Sometimes, however, the design team has to use company-internal personnel for confidentiality reasons, or when wider sampling is impossible because of tight schedules.

During a new mobile phone UI concept creation project at Nokia, which typically lasts about a year, we carry out three to four rounds of testing. When the design team is satisfied with the concept and it's getting close to implementation level, a design verification world tour is arranged. These are typically conducted in several countries, including a selection of the following: East and West coasts of the United States, Japan, Hong Kong, mainland China, and a few European countries. Rarely does a tour reveal anything radical; radical problems will have been ironed out before the designers packed the suitcase and set out. Regardless of their culture, people tend to approach logical usability problems and goal-oriented tasks much the same way throughout the world. Cultural differences have more to do with preferences in graphics and vocabulary than with the core interaction. Therefore the logical model of dialogue can be global, whereas displays designed for Europe and Japan have to be extremely different.

Usability tests, even when carried out with all necessary care, can be biased. There are several possible sources of error, including subjects' skills, subjects' motivation, the functionality of a simulation, and the inevitable differences between a laboratory environment and a natural one. Sometimes people involved in testing may interpret the results too blindly and propose a seemingly working solution that nevertheless has poor usability. The temptation to use usability as a lever for personal opinions and preferences has to be avoided by scrupulously critical interpretation of results, because biased usability is one of the world's worst sales tools. If it happens to a designer even once, his or her credibility will be destroyed forever. Think of marketing hype, where every product with a few keys and a display is pitched as "easy to use" even as customers struggle with it. Ease of use can be claimed very easily, and it is hard to get caught claiming it falsely. At Nokia, we do regard our products as usable, but for just that reason we seldom cite usability as a key selling argument in marketing communications. We prefer to maintain our credibility by letting that message be conveyed through word of mouth from user to user.

When describing new concepts, the designers must *always* be more specific than saying that a concept is usable. For instance, one can sharpen the point just by saying that the design is "more efficient" or "easier to learn" than another one. Jakob Nielsen's well-known usability attributes have turned out to be straightforward and practical tools for focusing usability. In a way they can be seen as usability dials for adjusting user experience with a given concept; we want 20 percent more learnability and 30 percent more memorability, and we'll trade you 30 percent of efficiency to get it.

Documentation Management

Tacit information is a stimulator for innovation. The fewer rules and conventions there are, the fewer restrictions for designers. With every frozen practice and fixed design guideline, an element of creativity disappears. Knowing that, we've been reluctant to document the Nokia user interface style, relying instead on tacit guidelines as long as possible. The tacit guideline is an implicit agreement about the look and feel of Nokia user interfaces, used to channel new design in the proper direction. Although it is sometimes criticized as being a synonym for "my manager doesn't like my design," it is more than that. It is more than a subjective opinion. A tacit guideline has a life of its own. It can be passed on as the responsible people change. It can be adapted, developed, and kept fresh. Even though Nokia is very engineering-focused and process-oriented, we did not document our core user interface guidelines until intense growth phase of Nokia. Gradually it became evident that the senior UI designers kept on answering the same basic questions. At that point, two senior designers, Johanna Järnström and Seppo Helle, gathered the principles together and gave them explicit expression.

The UI design guidelines are now like a mooring buoy in a natural harbor, saying that this cannot be taken away; we have learned something. We have a heritage. The fact that we have written guidelines has not changed our behavior. There is still a tacit thing called "Nokia style," and we think there always will be as a part of our culture.

The scope of UI documentation is naturally much wider than the design guidelines alone. Our UI documentation has a top-down layered structure. The highest-level document governing what we aim at is a "user interface vision and strategy." It covers major customer trends, technology progress and breakthroughs occurring outside the company, and the behavior of related industries before eventually setting forth what we wish to achieve, what actions we will take, and the schedules we will meet. (We call the latter a *UI roadmap.*)

One level of iteration from the roadmap, we find the *UI portfolio,* which describes key characteristics of each UI style (see Chap. 1). The next level of documentation is a style-specific *style guide,* such as the Series 40 or Series 60 style guide. These style guides describe the core functionality of the most important applications. They illustrate layouts and suggest how to design applications for the given style. The style guide naturally cannot answer all possible questions about applying the style. To reveal what the guideline does not know, we have nominated UI designers to act as the style "owners," whose job is to take the lead in style evolution and to assist in applying the style to new products or applications.

In product development projects, style guides are used to create *UI specifications.* These are very detailed documents defining the user interaction with a specific phone application or functionality. A UI specification is split into a common part, which is shared by all phones using the same style; and a product-specific part. UI specifications for a contemporary phone can be well over a thousand pages and written by tens of people. Fortunately we do not need to rewrite everything for each phone model. We can reuse all the common parts and thus speed up development and build more consistent user experiences. By reusing the common UI design, a new product variant can be designed in roughly one-third of the time and with one-third of the designers that were needed in creating the initial design.

To summarize, Nokia user interface documentation is a construct made out of

- The *Nokia UI design guidelines* explicating the Nokia user interface design heritage
- *UI vision and strategy* presenting a roadmap of product and style development
- *UI style portfolio* cataloging the UI styles that are at the disposal of product development projects

- *UI style guidelines* defining individual UI styles
- *UI specifications* defining the user interaction with an individual application down to the smallest detail

Consistency Management

Consistency is the best way to stimulate learning and create user comfort. It is also one of the hardest elements in interaction design. It would be relatively easy to make two solutions consistent with each other, but the reality is much more complicated than that. In fact, there are four types of consistency that have to be managed in parallel.

> *Intra-application consistency.* From a device point of view, the first aspect of consistency is *intra-application,* which means that all user-related elements of an application should be internally consistent. Let's say that the application is a calendar. To be internally consistent, the steps for booking a meeting should follow the same logic as the steps for making a to-do list entry.
>
> *Interapplication consistency.* The second aspect of consistency is that between applications. *Interapplication* consistency means that creating a note on a notepad should not be different from creating a note in the calendar.
>
> *Intrageneration consistency.* The third aspect is intrageneration consistency within the user interfaces of the product range. At any given time a Nokia phone model in one market segment should resemble models in other segments to the extent that the style segmentation allows. *Intrageneration* consistency creates challenges for the scalability of solutions. Using the same graphic metaphor for similar things will signal interoperability to the user, so if both products carry an envelope icon adjacent to a messaging application, it is likely that they can communicate with each other. However, many mobile phone displays have so few pixels that solutions that would be applicable in larger resolutions simply do not work. For example, a postcard symbol was used for text messaging in the first two Nokia communicators, but

not in the smaller phones—it just wasn't possible to design a postcard icon six pixels high.

Intergeneration consistency. The fourth type of consistency is *intergenerational.* The industry average for consumers upgrading their phones is 18 months. Since the typical phone model lifecycle in the industry is about 12 to 14 months, this means that if they replace their old phone with one belonging to the same category, they will be buying every second model. Suppose that the new phone behaves in a different way. The reason must always be a clear improvement that the user learns to appreciate. Change for change's sake generates negative consumer reactions. Obviously, intergeneration consistency may conflict with technology evolution, not to mention revolution.

The Management Toolkit

This chapter has introduced five dimensions of user interface management. These dimensions collectively form a menu of approaches that we think are applicable to any business in which smart product interfaces are developed. The details of how those tools are applied may need to be adjusted to the scope and dynamics of the business in question, but the selection has to be there at a minimum. To summarize, the approaches are:

Vision management. By painting attractive goals for the design, vision management drives development. It aims at lighting the sparks of innovation, getting the audience committed to the concept, and allowing everyone to contribute.

Team management. The skills of team management facilitate the birth of new creative solutions, concepts, and innovations by gathering together the right kind of competencies, motivating the group, and creating the conditions to foster its work. Competitions boost the imagination and ensure a scope of design proposals sufficiently wide for consolidating concepts.

Quality management. A continuous and gradually more focused usability evaluation program is a necessary management tool to secure the quality of the design. Managing usability assessment is about ongoing tradeoffs between keeping tight schedules and getting valid results.

Documentation management. User interface design generates and uses large amounts of information, the character of which varies from detailed specifications to overall insights to the company design heritage. These all have to be documented, and documents have to be updated, with their utilization supported. Planning and running the documentation system is a necessary, if perhaps not so exciting, area of UI management.

Consistency management. Multifaceted activity prevails in consistency management. Phone applications have to be internally and externally consistent. Solutions over the product range have to be as harmonious as segmentation requirements allow. Succeeding product generations may introduce different UI styles and solutions only when these novelties produce obvious end-user value.

In our research labs and in the labs of our suppliers, we find ourselves looking at the evolving and emerging UI technologies: displays, control devices such as micro joysticks, sophisticated browsers, and over-the-air downloadable applications. The next revolution may be where the terminal evolves from an impersonal object to an intimate possession containing one's most important data and thoughts. No matter how the mobile device interaction paradigms and technologies evolve, reliable management tools that have been hardened in a crucible of daily use will remain essential.

Miika Silfverberg

CHAPTER 7

The One-Row Keyboard
A Case Study in Mobile Text Input

By 1998 the short message service (SMS) had become a real hit in mobile GSM phones. The idea of transmitting 160-character textual messages to mobile terminals, originally meant for sending operator announcements, had taken off by itself—especially among young people. There was also rising interest in full-scale mobile email (electronic mail). Other kinds of text-based mobile services started to emerge as well; mobile phones were not just for talking anymore. These trends obviously increased the amount of text that the average user was entering into mobile terminals. Text entry methods existed, but they all had clear limitations. They were either too slow, too difficult to learn, or simply too large. Physical size is a major factor in mobile products; to be competitive, a product should be as small as possible. It should ideally fit into a user's pocket. A full QWERTY (the first six top left keys) keyboard with tens of keys is familiar to many users, but it takes up a lot of space even when it is scaled down.

A group of designers and researchers were assembled to find new solutions for mobile text input. Their goal was to generate concepts for several new input methods that would be nearly as efficient as a full-size keyboard in significantly less space.

The Idea—August 1998

By August 1998, an innovative text input method called *T9* for mobile phones had emerged. It had been created by a small U.S. company called

Tegic, today owned by America Online (AOL). T9 is based on standardized ITU-T* 12-key telephone keypad. T9 generated a lot of interest from mobile phone manufacturers, since it was the first major leap in mobile phone text entry from the traditional multi-tap (see Fig. 7.1) method. It slashed the number of key presses demanded of a user, potentially making text input much faster and less burdening. The results from usability tests with T9 prototypes at Nokia were largely positive, and the first T9-equipped Nokia phone, model 7110, was introduced in 1999.

Despite the benefits of T9, the 12-key phone keypad was beginning to be regarded as an unwarranted restriction in itself. The layout was fixed and relatively large, forcing the physical form factor of the phone into a conventional monoblock. This type of device positions the display on the top, the function keys in the middle, and the standard numeric keypad at the bottom. Emerging digital services, however, required us to study alternative screen size and orientation solutions where a rigid 3×4 keypad layout would have been a burden.

During August 1998, the team worked with several keypad layouts. The design driver was to find a solution that was as compact as possible, while still being easy to learn and use. The goal was *not* to replace the mobile phone keypad altogether, but to propose an alternative layout for mobile devices with larger display than the current norm. Potential targets were PDA devices, mobile Web browsers, and electronic notepads. An optimal solution would not compel users to learn a totally new system, but would build on some existing text entry method to allow a smooth transition from the old input style to the new one.

In one of the most appealing concepts, keys were arranged in a single row. This was promptly dubbed the "one-row keyboard" (Fig. 7.2). Character grouping was borrowed directly from the mobile phone keypad, but instead of a 3×4 matrix, keys were placed horizontally, side by side. The keypad was located right below the display. This positioning allowed us to utilize the idea of *softkeys*, where keys themselves have no label, but

---

*ITU-T is the International Telecommunications Union—Telecommunications Standardization Sector. ITU Recommendation E.161 (02/01), *Arrangement of Digits, Letters and Symbols on Telephones and Other Devices that Can Be Used for Gaining Access to a Telephone Network* (to be published); see *http://www.itu.int/itudoc/itu-t/rec/e/e161.html*).

their function is shown on the adjoining screen. This solution is very flexible because it allows several different functions to be associated with a single key, depending on the context, while clearly indicating changes of function to the user. Thus the keys in a one-row keyboard are not necessarily dedicated to text entry, but can be and were planned to serve several purposes.

As in T9, the keys on the one-row board are pressed only once for each character. The software, which can be identical to T9 software, proposes a suggestion for the intended word after a sufficient amount of characters have been inserted. Every now and then the user needs to correct the proposal that the software has generated, but with a good disambiguation algorithm the number of key presses per produced character is very close to one. Moreover, the layout accommodated efficient two-handed use. These factors together were expected to permit a very high typing speed, expected to approximate that of a full-size PC keyboard.

Entering Text with Mobile Phone Keypads: The Multi-tap, T9, and One-Row Methods

The standard mobile phone keypad (Fig. 7.1) contains 12 keys. Characters A to Z are spread over keys 2 to 9. There are three or four characters per key. In addition, there can be hidden characters beyond the printed ones that are also inserted with the same key. In the traditional *multi-tap* method, each key is pressed one or several times in succession. For example, number key 2 is pressed once for character A, twice for B, and 3 times for C.

Figure 7.1 The ITU-T standard telephone keypad (model Nokia 5110).

Disambiguation methods, such as the T9, allow users to press each key only once to insert one character. They use a dictionary to deduce the word the user intends. The most probable match in the dictionary is shown by default. If the word is not the one that the user has intended, the user can press a NEXT key, which is typically the star key (asterisk, *), to get the next most probable match. For example, key sequence 8–4–3 gives the most probable word for that sequence, namely, the word "the." However, to write "tie," the user needs to press NEXT key once. The one-row keyboard is similar to that of T9, but keys are arranged in a row. Two-handed use is possible. The keys are softkeys, without any printed labels. Instead, the labels are shown on the display. There is no NEXT key; instead, the matching words are scrolled through by pressing the SPACE key repeatedly (Fig. 7.2).

Figure 7.2 Illustration of the one-row keyboard.

High Hopes—September 1998

The one-row keypad was seen as a very promising input method. Many potential benefits were identified:

- *Small size.* Containing only a few keys arranged economically in a single row, the keyboard occupies very little space. The keys can be placed along the display edge, making the whole device only a little larger than the display.
- *Easy to learn.* Most people know the mobile phone keypad layout. Since the character grouping (2abc, 3def, etc.) is retained, the one-row keyboard layout would be comfortable for these users. The alphabetic order of characters is also familiar.
- *Fast two-hand input.* Although the one-row keyboard works to some extent with one hand, its benefits are most obvious when it is placed on a table or other flat surface. In these cases, the keypad can readily be operated with two hands. Even touch-typing would be possible!

- *Advantages over touch-screen input methods.* Although devices with a one-row keyboard could also incorporate touch screens, physical push buttons have clear benefits. The keys provide an immediate tactile feedback, and can be easily used with the fingers, not requiring users to carry and find a stylus. Most touch screens can also be used with fingers, but less easily, and grease in the skin smudges the display.
- *Softkey benefits.* Having the key label on the screen makes it possible to change the key function when needed. This creates flexibility in the user interface. It also makes text input easier, since the key labels are never covered by the user's fingers.

It was the last benefit, the flexibility of the softkey solution, that motivated the team to design a whole new user interface (UI) concept around the one-row keyboard. The concept included a relatively large touch screen. Text and data entry could be done either via the touch screen or the one-row keyboard, or both. In situations where the one-row keys were not needed for text entry, they remained functional as, for example, shortcuts to frequently used functions.

Preliminary usability testing was carried out with the new user interface using paper prototypes, and the results were encouraging. Besides the benefits mentioned above, some risks were also identified. For example, some team members came away with doubts about how easily the new layout could be learned. The risks were seen as small, though, compared with the long list of potential benefits, and the team was still excited about the idea.

From the very beginning, it was obvious that the *width* of the keyboard would be critical to the success of the whole concept. In a phone keypad, there are 15 keys that are used for text entry (the number keys 0 to 9, *, #, CLEAR, and SPACE, plus two arrow keys for moving the cursor). As the new keyboard was targeted primarily at touch-typing, the optimal key spacing would be about 19 mm (¾ in), the measure typically used in full-size desktop keyboards. With 15 keys in a row, this spacing would increase the total width to 285 mm (11¼ in), which is clearly too large for a mobile, pocketable device. [The largest contemporary mobile phones were around 150 mm (6 in) long at the time.]

The team decided to cut out some keys; only the most important functions would now have dedicated keys. Less frequent functions such as symbol input could be removed from the keyboard and operated using the touch screen. It was decided that SPACE and CLEAR keys should be retained, based on their frequency of use and importance. The so-called NEXT function of T9, which brings up the next most probable word match, was combined with the SPACE key, such that pressing SPACE repeatedly would scroll down the list of matches. These modifications cut the number of keys to 10. Still, with full 19-mm key spacing the keyboard was too wide. The hardware engineer of the group acquired several keyboards with different reduced key spacings for informal trials and comparisons. Finally, the team settled on 16 mm (⅝ in) key spacing. With these design changes, the width of the keyboard was reduced from the initial 285 mm (11¼ in) down to 160 mm (6¼ in). The size was regarded as a compromise. Touch-typing is not very convenient with key-spacing this narrow, especially for large fingers. On the other hand, total width was still quite large for a mobile keyboard. However, the compromise was good enough to begin user tests.

To facilitate the tests, a quick prototype was constructed. An old Toshiba 610 CT notebook computer was modified for the purpose. The function keys F1 to F10 were allocated as the 10 keys of the one-row keyboard (see Fig. 7.3). All other keys were removed. Pieces of paper were pasted above the keys to present the labels. The prototype was kept very simple, taking less than a day to build. It was reasonably good for the first prototype nonetheless, and key spacing matched the required 16 mm exactly. In a matter of days, the programmer in the project was able to come up with a simple Windows text entry demo for the prototype. We were now ready to face the reactions of the users.

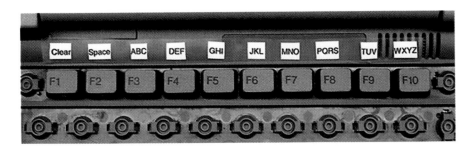

Figure 7.3 *The first one-row prototype. The keypad is designed for touch-typing, having a 16-mm key pitch. The total width of the keypad is 160 mm.*

Changing Horses—October to December 1998

The first evaluations took place during October and December. We wanted to get the tests started as soon as possible, so we took some short-cuts. Instead of using a large, carefully selected user sample, we ran the tests with just two of our colleagues. Also, the weight of our prototype did not allow handheld use, which limited the validity of the results. However, by making these compromises we were able to get some feedback at a very early stage of development. It turned out to be extremely useful.

At this stage the team was primarily interested in how users learn to touch-type. Therefore, instead of running several short tests, we ran a longitudinal experiment with fewer subjects. Subjects used the prototype one-row keyboard for a full 4 weeks, writing about 20 short texts each day. In total, they wrote over 24,000 characters during the experiment (which is very close to the amount of characters in this chapter).

Both subjects were regular mobile phone keypad users, sending tens of text messages per day. They had also been experimenting with T9, but didn't use it on a daily basis. Both used a PC keyboard daily and were relatively fast typists, using all 10 fingers for typing. However, neither could do real touch-typing. The subjects were instructed to "touch-type" throughout the test by using four fingers of both hands, each finger occupying one key on the keyboard. SPACE and CLEAR were operated by moving the left little finger off the ABC key. Thumbs were held idle.

During the evaluation, it became obvious that touch-typing with the one-row keyboard was not as easy to learn as we had hoped and expected. Typing with one-row was very slow in the beginning, since learning the new key locations took a while. For quite a long time subjects had to visually "hunt and peck" each key. The final typing speed was also lower than we anticipated; at the end of the test both subjects had attained less than half of their PC keyboard speed.

The most remarkable finding, however, was the vulnerability of the touch-typing ability. All went well enough as long as the subject was well rested, alert, and typing in an undisturbed environment. However, at the most minor disturbing factor, performance collapsed. Both of our subjects explicitly mentioned that typing was burdensome, and that they had to concentrate on it all the time to get the job done with any accuracy. The human factors expert in our group had a name for the burden; he explained *cognitive load* and *working memory capacity*. It was obvious that touch-typing took up the subjects' entire working memory capacity— or at least most of it, and then failed when other loading factors appeared. This was a huge problem, even a risk factor in the mobile context. We obviously did not want to see people running into accidents because they were using all their cognitive capacities to operate our fine new keyboard.

This result led the team to an obvious, but painful conclusion: *get rid of touch-typing!* The project changed directions, and went after what we called "four-finger typing" instead. It is similar to what many untrained typists do on their PC keyboards; instead of real touch-typing, they use a couple of fingers—two, four, or more depending on their expertise—in a relatively freeform manner. Four-finger typing seemed quite easy to learn at least for users with one-row experience. Our two subjects were able to reach their touch-typing speed in a matter of minutes.

Getting rid of touch-typing gave us one additional advantage. We could further reduce key width without compromising usability too much, as the fingers no longer had to fit side by side on the keys. Once again, the hardware designer went out looking for different key samples and key widths. This time we wanted to make the prototype more realistic, and issues like key "travel" and tactile feel were considered. The result was a prototype realized with small switches with exceptionally low travel and a very snappy tactile feel. The key spacing was reduced to 12.5 mm (½ in), a size that should be comfortable for four-finger typing. The full width of the keyboard was now reduced from 160 mm (6¼ in) to 134 mm (5¼ in) (Fig. 7.4).

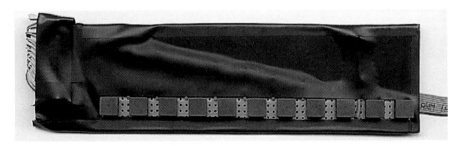

Figure 7.4 *The second one-row prototype. We give up on the idea of touch-typing and reduce the key pitch to 12.5 mm. The total width is now 134 mm.*

The whole device would be somewhat bigger than a plain mobile phone, but still be small enough to be carried around in a pocket. The team was now ready for the second user test.

Facing the Facts—January to March 1999

The second user test took place during the first 2 months of 1999. During this round of evaluation we made fewer compromises. The number of subjects was increased to eight. We were also careful to select subjects that better represented the intended user group. Yet some compromises had to be made. We were forced to use internal subjects, namely, employees of Nokia, since the one-row concept was regarded as highly confidential. Also, our prototype was still not handheld, so we were not able to test actual mobile use. However, we proved the value of early testing and decided to do it again. And again the testing turned out to be very worthwhile.

Eight subjects used the one-row keyboard for about one week. They were allowed to type in a free style, using as many fingers as they wanted. Since the keyboard prototype was now pretty small, all the users ended up using some variation of the four-finger strategy, as we expected. All subjects were regular phone keypad users. Earlier they had participated in a similar evaluation with T9, though none was a regular T9 user. That test was now used as a benchmark.

As users were familiar with the phone keypad, if not with one-row keyboard, we expected them to be pretty slow in the beginning. We measured their typing speed throughout the test, and expected it to surpass T9 speeds after some time. This hypothesis was based on the fact that regular T9 is operated mainly with one hand, whereas one-row is operated with two-handed input. We surmised that two hands would turn out to be more efficient in the long run. The hypothesis turned out to be false.

Figure 7.5 shows the typing speed learning curves of T9 and the one-row keyboard. One-row started from a lower level, as expected, but contrary to expectations the curves never crossed. In fact, they seem to diverge a bit. This finding was supported by subjective findings. Several users complained that the one-row keyboard was slow and tedious to use.

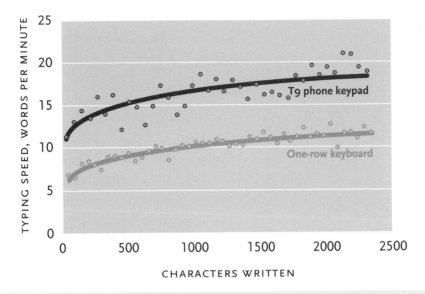

Figure 7.5 Typing speed during the evaluation with regular T9 and with the one-row keyboard.

After careful analysis of the results, there was only one possible conclusion: *the difference is real.* All subjects were consistently faster with regular T9 than with one-row, and there were no exceptions. Subjects' spontaneous comments about one-row were clearly critical. The team was stunned by the result; the two keyboards were based on the same operating principle. As one of them was used mostly with a thumb, the one with four-finger options was supposed to be an *improved* version. How could we have such a result?

For a while team members still tried to challenge the result of the user test: "C'mon, we ran the tests with only eight subjects. . . ." Some even took the trouble to learn to use the keyboard themselves, but were finally driven to the same conclusion. The keyboard was neither easy nor fast, and we had no choice but to face the facts.

Looking Back—April 1999

In order to learn from our mistakes, we wanted to analyze the results throughout. One observation that really stood out was the notion of "burden" or "load." Many subjects complained that they had to "think all the time" and "search for the keys." We had started with the knowledge that most users are familiar with the character grouping in a standard phone keypad, and assumed that this comfort level would be conducive to learning the one-row layout. Looking back after our two user tests, it seemed clear that this hypothesis was perhaps not altogether false, but at least overly optimistic.

We realized that the subjects often spoke in spatial terms. They would describe a certain key as being "on the left," or in a "northeast" direction. Some users had become very quick with common sequences, for instance, short articles like "the" or frequent word endings such as "-ing." They were typed almost automatically. This strongly indicated to us that the users' mobile phone keypad skills were more "in the muscles" than "in the eyes." So, instead of constantly gazing at the keys, skilled phone keypad users make use of their physical, procedural abilities in locating keys.

Another reason was also identified, also having to do with the spatial aspect. Whereas the standard 12-key phone keypad is two-dimensional, consisting of four rows and three columns, the one-row uses only one dimension, the horizontal one. This sounds like a small change at first glance, but it makes a big difference. In the phone keypad, the location of each key can be unambiguously identified by its orientation relative to a fixed origin. For example, if we take the middle key, the number key 5, as an anchor point, then each key has a unique orientation relative to this point. In the one-row keyboard this principle cannot be applied—all the keys are just "more or less to the left or right." The one-row keyboard, therefore, allows users to leverage their spatial ability only to locate the keys' *approximate* position; the final search has to be done using the visual domain. Therefore, although we had probably made users' lives a bit easier by retaining the character grouping, we had broken a connection that was even more essential: location in two dimensions. The character D that used to be "in the northeast" was not there anymore.

This second reason is a serious concern, since it might mean that the one-row keyboard is *inherently* slower than the phone keypad. Whereas some very heavy users of mobile phones can enter text messages even with their eyes closed, they might not be able to perform this feat with the one-row keyboard, which requires the additional step of visual search.

Although the one-row keyboard can be used with several fingers, each successive key press has to be *verified* with a single pair of eyes!

The issues mentioned here are fundamental—so fundamental that, although we could identify them, there was nothing we could do about them without abandoning the core idea. The idea was dead. The one-row keyboard would never be a marketable product. The text input issue was so central to the concept that the user interface built around the one-row format was invalidated along with it. The design team started looking at new ideas.

Looking Forward—May 1999

For some time spirits on the team were down. The one-row concept had been around for almost a year, and team members who devoted their passion to it, making new versions and prototypes, running user tests, and trying to improve it, finally had to let it go.

However, the question uppermost in team members' minds changed quite quickly from "What went wrong?" to "What's next?" All in all, it had been an excellent learning experience. In fact, asking what went wrong did not feel quite right, since many things had gone exactly the way they should. For instance, the team didn't just trust its own—false—intuitions blindly, but involved users in the design process early on. On the basis of user tests, the initial idea could be taken several steps further. It was only on completing the second usability test that we knew that we had hit a wall and ended the effort. We received invaluable feedback in the user tests. Observing and interviewing subjects gave us completely new and fresh insights. The lessons we learned with the one-row keyboard also helped us analyze other existing text input methods.

Mobile text input still remains a big challenge. All current input methods have serious limitations in the mobile context. The one-row experience helped us to see those limitations and come up with new solutions.

Pekka Ketola

CHAPTER 8

Series 60 Voice Mailbox in Rome
A Case Study of a Paper Prototyping Tour

Finland is a small language area, and the Finns' attitudes towards technology are exceptionally favorable. Since the early 1980s, patterns of social communication have been characterized by rapid urbanization. These are advantages for a communication technology company originating in Finland and still maintaining a remarkable percentage of its research and development resources in Finland. One's own backyard can be used as a test laboratory for future solutions.

Often, however, we find ourselves designing solutions for the real-world users of today instead of technophiles of the future. In these cases we have to speak with consumers where the markets are, where the customers live. That discussion has to be fluent and productive to satisfy the efficiency requirements of product development. It has to begin early in the development process for the users' words to exert their influence when it is most needed—when the solutions are not yet fixed. Consequently, we assess our user interface (UI) design drafts internationally with very preliminary material. We of course cannot afford to test them in each and every country, but there are countries and cities that we regard as good locations to represent surrounding market areas. Rome is one of those places. Mobile communication culture in Italy is mature, and mobile phone penetration is high. Rome reflects the southern European and Mediterranean communication culture quite well. Besides, why pass up a chance to go to Rome?

This chapter tells the story of one out of numerous tests performed on the Series 60 UI. We were in the early phases of creating a new UI style for future smart phones. Smart phones are an emerging product category where communication, namely, voice calls and messaging, is still the main function, but where personal information management is fundamentally improved compared to conventional mobile phones. Smart phones have good calendars, versatile contact management properties, to-do lists, address books, and so on. They are solid platforms for imaging and gaming. This chapter describes usability evaluations carried out with a paper prototyping technique and what we learned about the technique in international settings.

This study shows that although results from a single quick and dirty kind of usability test are relatively incomplete, the tour as a whole can be very instructive. The quality of data captured can be substantially enhanced if investigators also pay attention to cultural issues outside the immediate test sessions, learn about language differences, and repeat the tests on several sites. In addition to its primary purpose of eliminating usability defects from the present design version, usability testing is an opportunity for the different stakeholders in product development to learn about user behavior on a more general level and to gain insight into new UI solutions to be applied in future designs.

Scissors and Imagination

Friday afternoon, Tampere, Finland. An office table is covered with a mess of tiny pieces of paper. The people sitting around the table are reading aloud from them with very strange pronunciations, and every now and then roaring with laughter. An outsider would assume that they're just goofing off, having fun, and getting into the weekend state of mind. But no, they are on a mission.

These four people are trying to make sense out of 82 small printed display screens in Italian. The Italian translations for screens, test tasks, user questionnaires, and nondisclosure agreements have just arrived. Nobody understands the language except for a couple of words here and there.

The group is in the midst of a training session for a usability test trip that will start later this afternoon. Each one will have an opportunity to act as a test moderator, an observer, and a test subject. Eventually they decode Italian terminology well enough to follow the test task sequences; the relationship between the Italian screenshots and the familiar English version gradually become clear. "Connesso," "rubrica," and "indietro" acquire their meanings. In the meantime new display pictures are still being created—the total number now exceeding 90 options lists are being updated, material is being duplicated, organized, and packed. Two team members head for the airport.

We had begun design of the prototype just a few weeks earlier by defining a set of critical use cases. These were selected so that they would lead the user into performing and managing typical navigation sequences, such as finding certain menu options or starting, ending, and switching between applications. The use cases also reflected our need to learn more about the localized communication terminology in different cultures, the most critical design solutions in UI graphics, and novel interaction concepts. The Series 60 was in an early design phase and open to improvements and new design ideas. We were not yet fine-tuning.

The definition of use cases was followed by the creation of a prototype body and a set of printed screenshots in English. The test was prepiloted twice, verifying that the set of screenshots covered all the most probable user actions. Prepiloting sessions were also used as training for the test teams.

The usability test team comprised five different shareholders in the product development group: a UI designer, a localization expert, a usability expert, a native (Italian) market research expert, and an interpreter. This team included people who most directly needed information and a good grasp of the tested functionality, and who could also directly influence the design. In other words, having first-hand access to the development process from test results to product design optimized the effectiveness and efficiency of the test.

The prototype covered the whole scope of the product's functionality on a surface level and was focused on detailed sequences for the features that we were especially interested in; thus it was a combination of horizontal and vertical prototyping (Fig. 8.1). Several features were interlinked for fluent task switching—users could switch directly from call management to the calendar application, for instance.

The number of UI prints was no less than 80 at this stage, a large amount of material to be organized. A handy solution was to use folders (Fig. 8.2) for storing postage stamps, where the prints were immediately visible, were logically organized, and could be browsed easily. The body of the prototype was constructed from a phone-shaped piece of cardboard (Fig. 8.3) according to the mobile handset form factor. Buttons and UI elements were printed on the cardboard. Menu options lists, which are an essential part of the UI solution, were printed on separate paper sheets.

Before starting the tests in each country, we spent a day on prototype localization. This was done in a workshop where a localization expert went through the screens with a usability expert and looked for translated texts for each screen and option list. Translations were implemented straightforwardly on the screen pictures using the layering technique of a

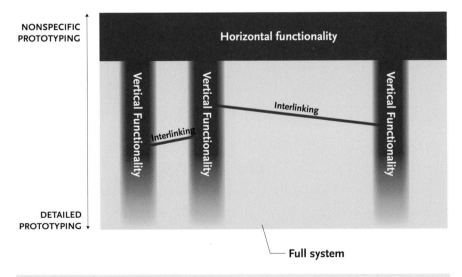

Figure 8.1 Horizontal and vertical functionality in a prototype. All features of the user interface are presented in a nonspecific manner. The most critical features are prototyped in more detail.

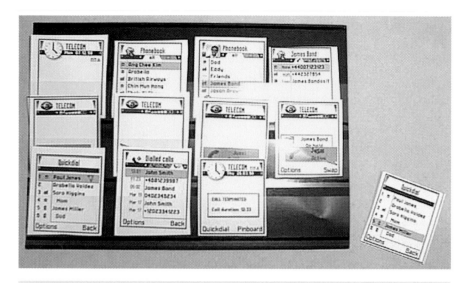

Figure 8.2 Prototype screen prints organized in a folder

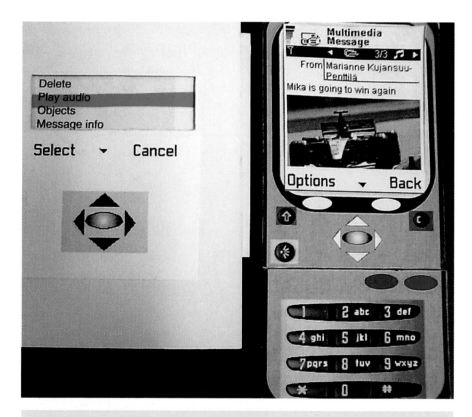

Figure 8.3 Cardboard prototype with color printing

graphics design application. This is a simple method for saving the localization information within each file, and it enabled easy printing of different language versions. The quality of the localized prototype was assessed for the final time during the kickoff sessions of the final test tour.

Message from Rome

Saturday afternoon I received a text message from Rome: "Lots of errors in the prototype and test material. Need to be fixed." Apparently the part of our test team that had flown to Rome to verify and check the test arrangements had done its job.

Once in the field, the usability test team was extended to include a native moderator, an observer, and an Italian-English interpreter. Saturday had been a training day. The vanguard of the test team trained the local staff to perform usability testing according to a prepared script, taught them our test objectives and got them familiar with our paper-based usability tools and made sure that everybody knew their role and responsibilities. During the training day it became evident—thanks to the local staff—that our terminology and test tasks were merely translated, not properly internationalized. Because of this some screenshots looked "funny" and some test tasks were described in an obscure and unnatural way. A phone conference was arranged, and the necessary changes were agreed on and implemented overnight.

On Sunday, the vanguard spent the whole day at the Formula 1 race at San Marino observing Italian mobile phone culture. The rest of the test team headed off to Rome from Finland.

We repeated this training-day schedule in each country of the test program, and it worked well for us. Apart from its project objectives, the training day also taught us details about language and local communication patterns. Test tasks were fitted to local culture, and localized test material got its final verification. A lot of language mistakes were corrected beforehand, and the suitability of the tasks for the local culture was

improved. For example, in Italy the daily use of voice mailbox—*casella vocale*—service was very different from the use cases anticipated, and we had to scramble to make modifications that would reflect local use. Regional facilitators thus assumed a secondary role as domain experts in the test preparations. It's not enough, we learned, for usability evaluators to be able to converse in a given language; they should also be familiar with the idiosyncrasies of the foreign language.[1] In a way, we ended by testing the user interface twice: first, by walking through it with local staff and again in administering the formal tests.

Two Hours Before . . .

> *Monday morning, 7 a.m. The team members meet at the hotel break-*
> *fast bar. Everybody is on edge about the upcoming test. By 9 a.m. we*

will be sitting at the test site on the premises of a market research agency. They have a large room for focus group discussions and interviews, a smaller room with audiovisual equipment for observers, and a wide one-way mirror between the two.

The first test user will arrive in 2 hours, and we have a long list of items to check. We need to clarify responsibilities and timetable once more, solve an unexpected technical problem, and pilot a short version of the test in order to confirm the last-minute corrections and overall test setup. When the first test user enters at 11 a.m., everything will be ready—barely.

During several rounds of usability tests we have noticed very similar patterns in the way things happen. On one hand we know that the schedule, originally planned to be unpressured, ends up being tight. Unexpected (typically technical) problems appear from nowhere; subjects become ill and cannot come. On the other hand, we observe that because usability testing is essentially dealing with people, and because people are very flexible and intelligent, things usually work out somehow. This is one reason usability testing is so rewarding and enjoyable!

I Am the CPU

The first test user comes in. Federica, our Italian moderator, gives a short introduction to the test situation and the test is under way. I, functioning as the central processing unit that operates the paper prototype, have a busy time finding correct displays and following the discussion between test user and moderator in Italian. This is something I hadn't anticipated—I couldn't hear the simultaneous translation from the observation room. This CPU is close to running short of MIPS.

The first adjustments to the prototype come early. A test user proposes better terms, and a missing function must be added to a menu list. Although I'm not following much of the discussion, I know that the team behind the mirror is listening to the translation and taking notes.

Testing with low-fidelity prototypes is a powerful tool in identifying interaction problems.[2,3] Paper prototyping is a technique where user interaction takes the form of printouts of screen designs. The printouts can be rough manual drawings on notepaper or elaborate graphics rendered with designer software. A human operator acting as a computer makes paper prototypes "functional." As each subject presses paper buttons and keys, the operator changes the user interface pictures and updates the user interface view to match.

Testing with paper prototypes provides several advantages. Speed in making simple simulations and modifications is a major benefit. In the best case it takes only minutes to create a simulation about a design idea. Applying the technique to competing design options is almost as fast. In our experience, among the other benefits of using paper prototypes are the extreme portability of the prototyping environment and, when needed, the ability to make design changes during the test sessions.

Consequently, paper prototypes provide an optimal approach for iterative design and testing. Repeating the tests not only increases the reliability of the results but also actually carries the design forward. The prototype itself evolves during the test period. Design and evaluation are not separate phases with this technique, but merge to form a user-centered design approach where idea creation, sharing, and assessment are intertwined (see Fig. 8.4).

> Tests can pinpoint areas for improvement in the following kinds of tasks:
> "How can I navigate from start view to calendar day view?"
> "Does the calendar day view provide all the needed functions?"
> "Where am I now?"

With a low-fi(delity) prototype test a user is likely to respond with suggestions about essential interaction issues, and hence the designer is able to see various modification possibilities. A highly polished high-fidelity prototype, on the contrary, fosters a sense of finality that tends to inspire only proposals for minor improvements and visual niceties.[4] So lo-fi paper prototyping is not just a compromise with tight schedules after all. By adjusting the level of realism in the stimulus in user tests, one is able to influence the level of detail on which the discussion takes place.

Figure 8.4 *Modifications in display pictures.*

Paper prototypes are the better choice for addressing major issues. The method allows instant design changes and thus immediate feedback to the user's concerns. In our study, instant design changes initiated by users included revisions to terminology and improvements to the on-screen instructions that the phone presents. The moderator and the person who operates the prototype are busy with their tasks, but the other members of the test team are able to produce design changes during the sessions and even during a single test task. They can participate directly and support the test situation in an interactive way. For mobile handset user interface design, this possibility has been a major source of innovation and simplification of interaction sequences. Some features, however, cannot be evaluated with low-fidelity prototypes. These are, for example, detailed graphics, audio input and feedback, real-time services, timing issues and quality of service (QoS)—all critical in shaping user satisfaction in mobile communication.

The relationship between the subjects and the paper prototype varies with the individual because using the prototype requires a certain amount of imagination and flexibility from the user. As a rule, though, as experience attests,[5,6] people tend to focus on the essential aspects of the paper prototype, namely, the tasks and interaction. As expected, the subjects in Rome were not paying attention to draft graphics and incomplete screen designs. Nor were they distracted by human error when, for example, the human CPU forgot what to do or displayed the wrong menu.

At the end of each test day the prototype was updated to reflect unexpected user actions that were encountered during the day and to implement design changes or new proposals that were too complicated to be done on the fly. Changes were carefully recorded for later validation. By the last test, the number of UI screens exceeded 100—large enough that the prototype could no longer be operated by anyone who wasn't intimately familiar with the system. At this point organization of the material became a critical factor in the dynamic test execution.

The Day After

What did we achieve? Did the product get better because of the test tour? Our concrete data consisted of a pile of display pictures with a lot of hand-drawn corrections and scribbled additions, video and audiotapes, and observation notes. Once we had combed through them, we agreed that the test engendered a number of insights, provided information about the acceptability of the new Series 60 user interface style, pinpointed specific design problems, and yielded several new design and localization ideas. The results are tabulated in Fig. 8.5.

Dumash and Redish[7] have created a classification of software usability problems. Figure 8.6 lists the main problems our tests uncovered and maps them to corresponding items in Dumash and Redish's classification. Their categories capture half of the problem types that we observed during our test round in four countries. The figure also shows that most of the observed problems could be fixed to final design.

As previously mentioned, paper prototyping is a tool to catch big fish in usability testing. However, the problems listed in Fig. 8.6 seem like relatively minor issues. Does this suggest that we were wrong to assume that

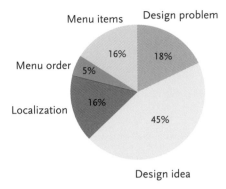

Figure 8.5 The distribution of different types of results in a paper prototyping test.

paper protos go to the very basic questions of a user interface style? Not necessarily. An excessive number of menu items, screen layouts that do not give hints about options not immediately displayed, and wasting time with data entry are either examples of problems that bring the whole style into question or just indicate unfinished design. It's up to the research team to analyze the implications. This time it was possible to fix the problems without having to scrap the basic core interaction logic.

The tests in Rome belonged to a series of four carried out in Finland, Italy, Germany, and the United Kingdom with 38 test subjects. The results from Rome confirmed, contradicted, or were neutral with respect to other findings in the research project as a whole. Several findings recurred in all test tasks and test sessions, and most of the usability problems were observed in at least two countries. For example, a specific but not obvious menu command that was not in the prototype was requested by a user in every country. This kind of observation can be regarded as a reliable conclusion and as a severe usability problem that has to be fixed. Contradictory findings were usually users' preferences concerning the order of menu items or comments about localization issues. Some problems were seen in only one country and they most often concerned a missing note or a text that was hard to understand, requiring either language-specific fixes or the addition of clarifying information.

OBSERVED PROBLEM	FI	ITA	GER	GB	DUMASH AND REDISH	FIXED
Option list is too long	●	●	●	●	Too many menu options	●
Requested menu option is missing	●	●	●	●		●
The order of menu items is wrong	○	○	○	●		●
Numbered menu items requested	○	●	○	○		—
Don't understand that there are more options than shown on screen	●	●	●	●	Don't understand that there are more options than shown on screen	●
Colors should link to user interface and physical buttons better	●	●	○	○	Function keys arbitrarily mapped to functions	—
Immediate editing should be possible in certain kinds of forms	○	●	●	○	Waste time entering the data	●
A note/expression/term is not understood	○	●	○	○	Error messages need additional level of detail	●
A confirmation note is missing	○	○	○	●	No confirmation that user intends action	●
An information note is missing	○	○	○	●	No message that action has been taken	●
Shortcuts for function are missing (several places)	●	●	●	●		●
Term not understood	○	●	○	○	Menu options use jargon	●
Term localization proposals	●	●	●	●		●
Functionality is available too deep in menu structure	○	●	●	●		●
Handling of data connection termination is not as excepted	○	○	●	●		●
Default functionality of selection key was not as expected	○	○	○	●		—

Figure 8.6 Observed usability problems and mapping to the Dumash–Redish classification.[7]

Voice Mailbox di Italia

When a new UI style is designed, it has to be assessed with real users who come from different cultural backgrounds. In this chapter I have suggested that it is possible to conduct usability evaluations with low-fidelity prototypes that support design for universal usability and allow culture-specific changes to be introduced during testing. This enables the fast design iterations that are necessary in intercultural modifications to our user interfaces. The simplicity of paper mock-ups prompts the test users to give feedback about important basic interaction decisions rather than design details.

By repeating the tests, we can gain the following advantages:

- The reliability of results increases.
- Iterative product design is supported in a natural way when changes are implemented immediately and evaluated in the next test an hour later.
- The test setup can be continuously improved by iteration. The first test is different from the last test even though the test plan is identical.

Figure 8.7 summarizes the phases of international paper prototype testing over the 1½-month period that was the total length of the test round. Steps from prototype localization (day before testing) until test sessions ended (day 4) were carried out in parallel in four countries. Although the prototype itself can be prepared quickly with a paper prototype, time is needed for practical arrangements, communication, reviewing, and figuring out challenges posed by the temporary project organization that this kind of testing represents. Figure 8.8 shows the approximate number of workdays required from the shareholders. The figures exclude the actual design work that precedes testing.

The key activities, with approximate overall workloads, are

Test project coordination: 30%
Documentation (including pretest preparations and posttest analysis): 30%
Execution of usability tests: 40%

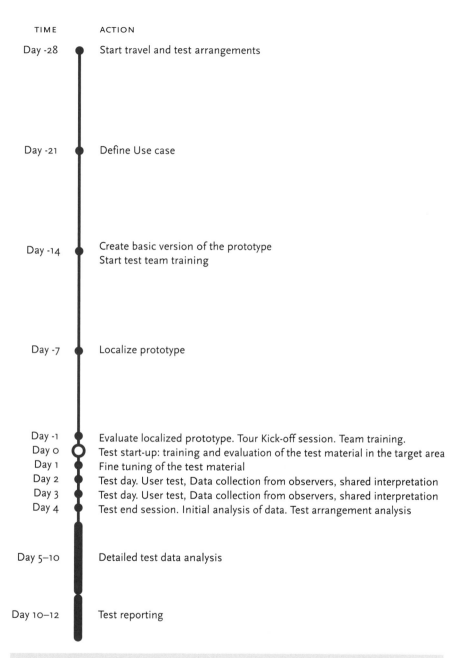

Figure 8.7 The process of preparing an international usability test round.

TASK	UI DESIGNER	LOCALIZATION EXPERT	USABILITY EXPERT	LOCAL MARKET RESEARCH EXPERT	TOTAL
Prepare test plans			1–2		1–2
Adjust the UI style specification for test	1–2	1–2			2–4
Prototype building and pilot testing	2–3	2–3	1–3		5–9
Arrange the test facilities, people, travel, timetables			5–10	1–3	6–13
Participate in the test	4–5	4–5	4–5	3–4	15–19
Detailed test analysis			5–10		5–10
Prepare test report			1–3		1–3
WHOLE TEST					**35–60**

Figure 8.8 The amount of work (expressed as working days) needed to arrange and conduct an international test tour.

This round of Series 60 UI evaluation just described was limited to Europe. We did other tours including the east and west coasts of the United States, Japan, and China. International test rounds are carried out for validation of major new interface solutions, like Series 60. Usability world tours address both the common core of usability and local differences. The reasons behind cultural variations are perhaps better revealed by more contextual approaches, as discussed in Part 2 of this book. Paper prototype testing, however, tightly links cultural differences in a concrete way with the UI solutions we're evaluating. One can say that Nokia as an organization has learned to excise designer-centered thinking, whereby something that is appropriate and understandable for me and the people I know is good for everyone else as well.

The Series 60 voice mailbox is presented here as an example of local differences in using mobile services. The rationale for our product design came from the Finnish way of using voice mailbox in a mobile handset, and from a designer-originated way of looking at the feature. In many handsets the voice mailbox is accessed with a shortcut in key 1, and in our operating environment one's 10-digit voice mailbox number is based on the user's personal mobile phone number. Manually dialing voice mailbox would require the user to insert all 10 digits before pressing the SEND key. No one does that; there is plenty of motivation to find a shortcut.

We assumed that the key 1 shortcut is generally used. In Italy, though, voice mailbox is accessed by dialing 919 instead of a shortcut key. In our tests most users decided to dial the number manually because they couldn't find a corresponding function directly indicated in the UI, and because they were accustomed to performing this operation that way. Therefore, if we want to improve voice mailbox access in the Italian context, cutting down the number of key presses is not enough. Something else needs to be done, for example, replacing the shortcut with a clearly labeled menu item.

We could have asked Italian operators how they provide their voice mailbox services, but by asking users instead of mobile operators, we learned not only what is provided but also what is used and how it is used.

References

1. A. Yeo, "Usability Evaluation in Malaysia: An Exploratory Study of Verbal Data," in *Proceedings of 4th Asia Pacific Computer Human Interaction Conference: APCHI 2000,* Singapore, Nov. 27–Dec. 1. Amsterdam: Elsevier, 2000.

2. S. Isensee and J. Rudd, *The Art of Rapid Prototyping.* London: International Thomson Computer Press, 1996.

3. H. Kiljander, "User Interface Prototyping Methods in Designing Mobile Handsets," in *Human-Computer Interaction INTERACT '99,* 1999, pp. 118–125.

4. T. Winograd, "From Programming Environments to Environments for Designing," *Commun. ACM* **38**(6): 65–74 (1995).

5. P. Ehn and M. Kyng, "Cardboard Computers: Mocking-it-up or Hands-on the Future," in *Design at Work,* J. Greenbaum and M. Kyng, eds. Hillsdale, N.J.: Lawrence Erlbaum, 1991.

6. S. Säde, M. Nieminen, and S. Riihiaho, "Testing Usability with 3D Paper Prototypes—Case Halton Systems," *Appl. Ergon.* **29**(1): 67–73 (1998).

7. J. S. Dumash and J. C. Redish, *A Practical Guide to Usability Testing.* Portland, Oreg.: Intellect Books, 1999.

CHAPTER 9

¼ Samara
Hardware Prototyping a Driving Simulator

Behind the cafeteria counter stands an inconspicuous entrance that leads to a long corridor. It is locked; this corridor is evidently inaccessible to outsiders. Several doors appear at intervals along its length, one of which leads to a stairwell. It, too, is locked, unmarked, nondescript. If you were looking for it, you'd be hard-pressed to recognize it. Enter it, and a flight of stairs leads you down to the windowless underground floors of the Research Center. The third locked door (counting from the cafeteria) opens into a basement passage that quickly turns into a small lobbylike space. On the right-hand wall is a steel door with an alarm. The door reads TeleRing, a mobile phone retailer. It's a bluff. It has nothing to do with TeleRing.

Two men from the Research Center are wandering around the Tattarisuo junkyard. Snow still hides some of the cars late into spring. The men open the doors of the cars when they can or just peer in through the windows. A dark blue Audi 100, probably a 1994, is in good condition considering its location, but it doesn't interest the men for long. A red Ford Sierra is quickly bypassed. So it goes, one car after another. The men had already passed the white Samara—a Russian-made family sedan—but they return. The left front fender is mangled, but the dashboard and seats are in good shape. The interior is austere, practical. The radio works. One man points out details in the dashboard and speaks rapidly. The other is skeptical, but after a while begins nodding his head. The two edge past a German shep-

herd guarding the office and step inside. "One-quarter of the white Samara," they say, "delivered."

Representatives of a leading carmaker are giving guests a tour of the product development department. After a glance at the mandatory organizational charts, they set out for the research facilities to look at laboratory equipment. A shiny new car just off the line has been parked in the lab. A closer look reveals that someone has replaced the instrument panel with a display terminal, and the controls are connected to a computer but not to the engine or transmission. This vehicle's cockpit faces a large screen on which a projected image of a flat highway in a rural landscape wavers. Loudspeakers are everywhere. There are no wires in sight. The racks and tables have been built of aluminum profiles. The space has that new-car smell. You can't help but notice that the guests are half-smiling, thinking about the basement back home in the lab.

Electronics play an ever-more-significant role in automobiles. New premium cars are equipped with diverse systems for music, radio, navigation, communication, and traffic information. Similar systems will show up in ordinary family cars soon, and to some extent they already have. The systems can be purchased and installed as separate aftermarket equipment, but the integration and factory installation of the systems is a growing trend.

Many of these car systems are rather complex from the usability angle, and combining several applications into one centralized user interface further complicates the matter. The automobile is a challenging environment in terms of ergonomics and usability, because its electronic devices are used by the person driving the car. Some manufacturers have restricted usage of car computers or navigator settings such that their features are functional only when the car is stationary, but in general car systems are intended for use in traffic. Drivers thus have to divide their attention between operating the car and the electronic systems inside it.

During the late 1990s no one gave much thought to what the user interface for an automotive electronics system would look like when all the subsystems under development, and those already partially or completely realized, were integrated. Car manufacturers mastered the driving part of their products, but electronics were bought directly from the equipment manufacturers. Each vendor focused only on its own limited area of expertise—the audio system, the navigator, the phone. Meanwhile, studies were under way to determine the traffic safety risks posed by these devices, particularly the phone, and authorities in different countries were imposing regulations that targeted individual features or functions. Prohibitions to using mobile phones while driving—without hands-free equipment or, in some cases, at all—were enacted.

What such regulations failed to acknowledge was that in many countries people spend long periods of time in their cars. Traffic is congested and moves slowly. Drivers want to take advantage of this time by getting a part of their work done or at least by entertaining themselves. They try to do as many of the same or similar types of things in their vehicles as they do elsewhere. For this reason the car's systems must seamlessly integrate with the information and communications technology they use in their offices, at home, or on the street.

For several years at the end of the millennium, a lot was happening at once in in-car communications; a variety of systems were being integrated for the first time, safety regulations were tightening, and drivers started to require improved compatibility between automotive, desktop, and handheld products.

The Steering Wheel Project

Nokia's car products on the market at the beginning of 1999 followed established solutions. They were standalone devices providing hands-free, battery-charging, and radio-mute features. The user interface of a handset was used during driving as well. New ideas had come into play to improve one aspect of these car kits or another, but there was also a conspicuous lack of direction for UI development in the automotive environment.

To create one, it was decided to launch a concept project called the "Steering Wheel." The goal was to create insight into the design of in-car user interfaces in a midrange timespan, specifically, to outline possible products to be launched in 5 years. Additional goals included finding concrete solutions that could be utilized for immediate product improvement, and increasing general know-how for the user interface and industrial design of integrated car systems. Steering Wheel focused on drivers' behaviour during driving, as well as the periods immediately before a trip starts and after it ends. Communications solutions naturally were foremost in Nokia's vision, but the underlying aim was to design communications to integrate seamlessly with the other systems in the car, and with the driver's information processing needs outside the car.

The year-long Steering Wheel project was launched at the beginning of 1999 and carried out in close cooperation with product marketing, industrial design, and usability organizations. It was divided into three main phases: collecting background information, parallel design and research operations, and integration and testing. Background information was collected with many methods at several points of attack. The project participants became familiar with traffic psychology. They mapped the car manufacturers' visions of future user interfaces and tested prototypes, commercial products, and integrated car systems. The most significant effort to acquire background information was a 6-week driver observation period conducted in London, Munich, and Salo. During these weeks, researchers actually traveled with 30 drivers and observed their actions during work commutes and business trips.

After the observation-oriented initial phase, design and research were launched in tandem. The main objectives were isolated and, to give the handling of those objectives more depth, several simultaneous interaction design and industrial design exercises were started. At the same time, a test environment was built, and design and user interface solutions were proposed along with interactive prototypes and models to solve the critical usability problems. From this point on the project focused on the design challenges around drivers' attention sharing between controlling the vehicle and operating in-car systems. Observations also underlined the vast individual differences in drivers' skills and confidence in using communication devices during driving. These framed the scope of design tasks much wider than we originally assumed.

In the final phase of the project, the best ideas that had emerged in the parallel design phase were incorporated into concept designs that matched and complemented the observed user needs.

Speed Does Matter

Time is a critical factor in the design of a car user interface. When using the car systems, drivers have to divide their attention between following traffic, controlling the vehicle, and using in-car equipment. Traffic psychology specialists have extensively studied the drivers' divided attention and its impact on driving performance. They have shown that the control of a car is affected by both how long the driver's eyes are away from the road and the total time of the secondary task, specifically, the operation of an in-car device.

There are big differences between individuals. Experienced drivers can control their cars using their peripheral vision much better than novice drivers can. On the other hand, it is impossible even for them to react quickly in emergencies if their eyes are directed away from the road. A general rule of 4 × 1.5 (preferably 3 × 1.2) seconds for the maximum duration of a secondary task has been suggested.[1] According to this rule, any task performed during driving must not require drivers to take their eyes off the road more than 4 times longer than 1.5 seconds at a time. The following figures give an idea about the distractive effects of some common tasks drivers carry out:[2]

ACTION	NUMBER OF GLANCES	LENGTH OF GLANCE, SECONDS
Reading a speedometer	1.26	0.62
Inserting a cassette	2.06	0.80
Manually adjusting a radio	6.91	1.10

These figures gave us a framework for designing and evaluating solutions for in-car secondary tasks such as finding entries in a contact directory, adjusting receiver volume, and browsing a calendar.

In terms of the Steering Wheel project's methods, the importance of the user's divided attention meant that UI solutions had to be evaluated right from the start in a dual-task environment. The primary task had to approximate driving with respect to its attention-catching requirements, and the test environment had to physically resemble the dashboard of a car so that the control elements could be placed in their "natural" places within the drivers' physical reach and line of sight.

How to test was initially a problem for the project. A real car couldn't be modified enough to meet the needs of each different test situation. And tests with unfinished prototypes would have been downright dangerous—not a defensible risk for a project intended to design safe traffic products. Car factories and research facilities have expensive and advanced simulation environments that can move the car's chassis dynamically so that the driver gets a natural sense of acceleration when speed fluctuates and during turns. This kind of simulated driving environment can be fine-tuned so that the driving experience closely resembles the reality. (In the best simulators the environment is projected onto a hemispherical shape that covers the driver's entire field of vision.) Hiring a test environment would have been an option if we had testproof prototypes ready for a compact evaluation session, but we didn't. We needed a simulation environment at our disposal throughout the project for small-scale experiments and iterations of the most interesting ideas. The simulation environment phase of Steering Wheel was budgeted for about one calendar month and had a limited amount of funding with which to create a simulation environment, so anything fancy was out of question.

High-Tech Samara

It was the junked Samara that saved us by becoming the chassis of Steering Wheel's simulator. The Samara is a Russian-made family sedan with a reputation for providing good value for the money in the cheapest category of cars. It is technologically very basic, and the design is plain, even clumsy. The junkyard agreed to saw off the engine compartment and the entire rear from the B pillar back. In the spring of 1999 the Samara carcass was delivered to the brand-new high-tech palace in Ruoholahti, Helsinki, where the Research Center had just moved. The original intention was to install the Samara in the usability team's offices. The mangled metal, however, bristling with protruding wires, did not conform to the style guidelines of the new information society—not even with the special wax job the junkyard threw in for free. After many acrimonious exchanges, a place for the Samara was found in the deepest recesses of the basement where nobody would catch sight of it—not even accidentally.

Samara was not chosen by coincidence. It suited the purpose because of its design and brand. The Samara dashboard has a very open design. The gauges are relatively small, and the dashboard is close to the window, far from the driver. It was easy to build a new, modifiable dashboard made of cardboard over it without having to completely dismantle the old one. The brand also had significance. While this project was under way, the possibility of future cooperation with the automotive industry was an open question, so we had to choose a brand that was neutral and could not imply a commitment to any manufacturer as a future development partner. Better yet, the Samara was ideally suited for the UI design's rough and dirty prototyping philosophy: *using cheap materials to build a deformed but working simulator.* With cutting and delivery, the Samara cost EUR 500.

Researcher and prototype expert Topi Kaaresoja and design apprentice Janne Kouri started modifying the Samara immediately. They removed the center of the steering wheel, the pedals, the gearstick, and other unnecessary elements. The steering wheel axle was then attached to a PlayStation game controller wheel. The pedals were replaced with PlayStation pedals. The original gearstick was swapped out in favor of a new one that connected to the game controller's gearshifter. The old dashboard supported a new one built of cardboard. It was covered with Velcro tape so that various equipment could be easily attached and relocated. The center of the steering wheel was also replaced with cardboard to house different types of switches.

At last the controls were connected to the game console and the console connected to a video projector placed first on the roof of the Samara, and later on a separate rack. The image was projected onto a wall 4 m in front of the Samara. This distance was deemed to be the shortest possible to allow a driver's eyes to refocus between the road and the instrument panel on the dashboard. Speakers were installed in the ceiling of the cab to carry game sounds—sound plays an enormous role in creating a genuine experience in a simulator environment. It actually makes the driving experience easier because sound conveys a good sense of the vehicle's speed and thus helps the driver instinctively respond to acceleration.

Finding a suitable driving game for the simulator was not a trivial task. Gran Tourismo, Monster Track Madness—luckily there were among us selfless researchers who tirelessly tested and compared games, tracks, and simulated vehicles for hours on end. There were games for PlayStation and driving simulators for PCs. Eventually, we settled on the Need for Speed for PlayStation. PC applications crashed, and restarting them was agonizingly slow. In most games, controlling the vehicle was excessively difficult. In Need for Speed, it became easy enough to control the car at slow speeds after practicing for a short while. But new problems arose when testing began: after the driver completed his first lap, the game, assuming that the driver had reached the intended goal, stopped and had to be rebooted. Were that to happen during a usability test session, task performance would be interrupted and test subjects would be distracted, interfering with results. Besides, we couldn't face starting the game over and over again during each session. Eventually the problem was solved by the expedient of making a U-turn on the track at the start. By driving the track in the opposite direction, the driver eluded the game's propensity to stop at the end of every lap—a perfectly impressive little innovation.

The next problem emerged during the first live tests. The members of the project group had driven the Samara a lot and were pleased and impressed with the video projector's big picture. The test subjects, however, surprised us by suffering from motion sickness: Need for Speed routes cross rather hilly terrain with plenty of turns. The wall-sized motion picture captured the mind but, unfortunately, also some stomachs. For the remaining tests, the projector had to be replaced with a TV monitor placed in front of the Samara.

To record the tests a video system with two cameras and a video mixer was built around the Samara. The video system was totally separate from the actual driving simulator. Depending on the test, we recorded the driving performance from the monitor, the operation of the keyboard interface, the display of the user interface, the direction of the driver's gaze, an overall view of the test situation, or some combination of these.

User interface simulations were captured by two videographics array (VGA)-resolution LCD color displays installed in the Samara—one in the instrument panel and the other on top of the center console. The displays were controlled with an industrial PC. The keys of the simulator and other controls were connected to the PC via a keystroke encoder. The moderator next to the car was equipped with a separate display and a mouse. All simulations were programmed using ToolBook and Delphi software packages.

Creating user interface simulations isn't a matter of implementing a chain of well-reasoned concepts. Instead, plans progress according to the prototyping. Prototyping always begins on the basis of an indefinite description; you can't expect to begin at the most natural starting point because that characteristic might not have been determined yet. While the plan is fluid, it will be impossible to hunt down the right components or to design the optimal software because anything can change. One interesting, new component can redefine the concept. From a designer's standpoint, the prototype isn't likely to change much, but all the work done to implement it might be wasted. Making a prototype is a continuous tradeoff of what can be frozen and what must be capable of adjustment. This kind of work is extremely stressful and frustrating. The unfinished prototype must constantly be available for testing and for different demonstration purposes. There is no time for applying finishing touches—at least not until the night before the test.

Because there is no time to spare, nothing can be built properly in a prototype. The solutions are made to function reasonably well and to be relatively durable. Everything in the Samara withstood the design team's tests very well. The first test user steered a little harder than anyone in the design team, and the garden hose connecting the steering axle and the game controller couldn't take it. Our response was to equip the axle with stoppers to prevent oversteering. Tests were put on hold for that.

One challenge of car UI design is to offer direct control to several functions, but in such a way that the user interface seems simple. A possible solution to this problem is to use multifunctional switches whereby one element can directly control many parameters. When we examined the usage logics of various solutions, five functional switches proved interesting. Sony car radios had such a switch—in the middle is a push button surrounded by a freely rotating ring, and a lever switch sits just outside the ring. The size of the element makes it easy to operate. All its parts are easy to adjust individually, but the totality looks clear and simple. The switch was not quite what we were looking for, but it was the closest

ready-made element we were able to find. We purchased a car radio, dis-
assembled it, and built the control element that we needed from the
switch. Basing our calculations on those for the first element, we sized the
control element's connection onto the panel of our finished prototype. The
concept required four similar elements, so we bought three more radios.
After disassembling them we noticed that parts that appeared identical
actually had internal dimensions different from those of the first one.
There was no time to make a new panel, and since we had sawed the
front panel into pieces, we doubted that the radios could be returned to
the store. So we threw budgets to the wind and purchased new radios,
this time being more careful to check the model markings. Our shelf now
held eight functionally totally useless radios. The Samara was now ready
for the track (Figs. 9.1 to 9.3).

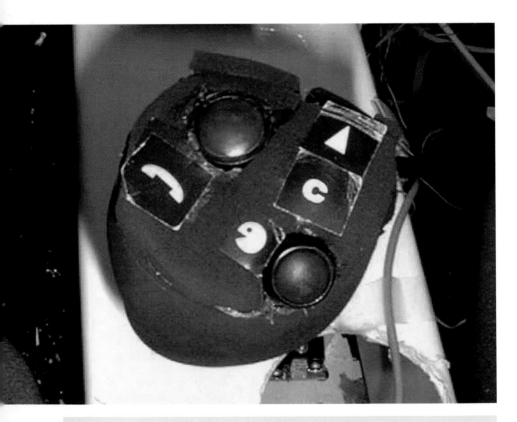

Figure 9.1   Modified Navi-key-style UI controls on the gearshift knob.

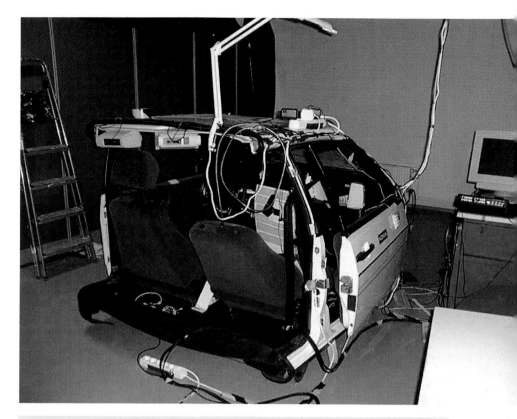

Figure 9.2 The Samara is ready for the track.

On the Track

While Steering Wheel ran its course, the Samara was used to conduct several tests. It was repeatedly equipped with different control elements for call handling. The positions of different physical control elements were assessed. Solutions based on voice recognition were compared to a user interface with buttons. Eventually a complete integrated UI simulation was built into the Samara. Let's look a bit closer at one of the tests conducted, designed to facilitate text entry during driving.

Writing while driving presents the designer with conflicting aims. It is obviously dangerous to write while driving a vehicle, so perhaps a solution encouraging it shouldn't be designed at all. On the other hand, some

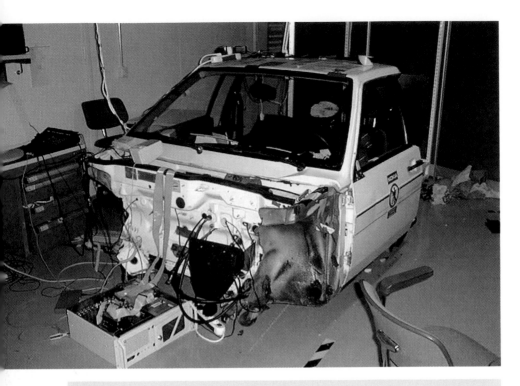

Figure 9.3 Another view of the Samara is ready for the track.

drivers write anyway. They write with devices that were not intended for use while driving, such as the keys on a mobile phone or a pen and a pad of paper. So perhaps it would be better to offer a solution after all, a solution optimized for use while driving. Text entry does not necessarily involve long messages. It is more often needed when, for example, recording a name or an address. It is also needed when searching for place names in the navigator's memory or for names of people in the mobile phone's phonebook. Keying in just a few letters can simplify and accelerate the search substantially. Without taking a stand on the acceptability of entering text while driving, we decided to conduct trials of what can be done so that writing needn't disrupt driving as much as it does with today's devices.

In one of these trials researcher Jarkko Ylikoski added voice feedback to the buttons of a normal ITU-T standard phone keypad. A synthesizer read the letter aloud as soon as it was entered. After a user keyed in the space button, the synthesizer would read the entire word. She could also replay the last three words entered by pushing the star button. Voice feedback realized this way supported writing on the level of individual letters. The spelling of a word could be checked after the whole word was completed, and listening to the last words entered helped recall the context if the message was interrupted by traffic. On the basis of this idea, we constructed a simulation whereby the phone was connected to a PC that produced the voice feedback, and installed it in Samara's dashboard (Fig. 9.4).

During the test, subjects were asked to drive and write simultaneously. The texts to be written were short, one- or two-word names and addresses

Figure 9.4 *The condition of Samara's cabin at the time of the writing test. The writing task was done using a Nokia 5110 phone, which was connected to a PC to produce voice feedback and to control the display. The buttons on the steering wheel and the LCD displays are in the cabin for other tests.*

with letters and numbers. The same tasks were performed with and without voice feedback (see Fig. 9.5). We registered no difference between the voice feedback and standard solutions as measured by the established quantitative usability criteria, specifically, the speed of the performance and the number of mistakes. In addition to the established usability criteria, we also considered the length of time that drivers were out of their lanes while performing the writing task. This measurement was used to assess the extent to which the handling of the phone interfered with the primary task of steering the Samara. The actual measuring was done after the test by analyzing videotapes. It indicated that voice feedback had a very significant effect on the drivers' ability to stay in their own lanes.

One problem yet remaining was that the driver had to visually focus on the keypad now and then in order to find the correct buttons to push. With

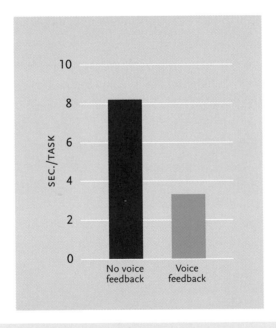

Figure 9.5 *The duration of straying from the driving lane during the writing task. The bar on the left indicates the duration of straying when the writing task is not supported with voice feedback, which is the situation with today's phones. The bar on the right indicates the situation where writing is supported with voice feedback on a letter level, word level, and sentence level. The duration of straying has decreased by more than half.*

a 3 × 4-keypad layout it is possible to use the keys solely on the basis of memory or by touch, but few phone users have learned to do that so well that they can do without visual supports. Simultaneous concentration on steering and writing increased the user's mental taskload to the point of discomfort. Thus the subjective evaluations about voice feedback were positive, but not as enthusiastic as we would have hoped according to the quantitative measurements.

At the Finish Line

It turned out that driving the Samara took some practice and was harder than steering a conventional car. However, it furnished a fair approximation of a natural driving situation and offered the physical environment of a real car interior. Our subjects were given the chance to get used to the simulator before they took up actual tasks. They were asked to find an individual driving speed that was as fast as they could go while still feeling in total control of the Samara, and then to maintain that speed throughout the tests. Some people had trouble getting used to the car, but most projected themselves into the driver's seat fairly easily. The subjects generally avoided collisions and were clearly sorry if they did crash. (The collision sounds in the game were particularly impressive.) After the test concluded, some people parked the car on the side of the road so that other traffic could pass.

Because of the unrealistic nature of the simulator, there was no intention to make any absolute measurements. All tests were done as benchmark tests with several new solutions compared to each other, or with one new solution compared to existing solutions.

Still, the Samara was not just a test environment. At least as important was the fact that the designers themselves could step into the Samara and try the different solutions. In this respect, the Samara functioned better than more sophisticated driving simulators. We now believe that the design environment must be available to the design team for a drive at any time to try a new solution or to recall attributes about old ones. This kind of ready availability is impossible with the expensive simulators; they cannot be reserved for one project's use for a long enough time, and they frown on drilling new holes into the design environment or messing it up with hot glue and tape. If an environment is going to accommodate new ideas, it must be open to modification.

An interactive prototype—demo, simulation—is the core of a UI design. It plays an important role in planning, testing, and communication. It forces a detailed specification of the interaction, enables the user's active participation in the process, and helps communicate ideas forward. Testing the simulated concepts is an essential part of the prototyping process, and the methods used in it evolve and take shape alongside the available concretizations.

There were plenty of shortcomings with the Samara. Measurements were not as controlled as they should have been. Drivers might have compensated their performances by adjusting their driving speed in ways that were difficult to take into consideration when analyzing the results. The game caused random incidents that might have influenced the results. The driving feel could have been improved so that the performance challenge corresponded more closely to real driving situations. Automatic recording of performances would have saved a lot of time in the results analysis phase. The simulation environment should have allowed for measuring drivers' reactions to a car braking in front of them; this was not possible with the Samara. The list goes on and on.

Nevertheless, the Samara fulfilled the expectations set for it. We were able to use it for preliminary testing, which was the target level in the Steering Wheel project. (Others took the ideas that seemed worth additional attention and continued to explore them.) Samara has also been used in other tests when the goal has been to research the use of a phone as a secondary task. It has been easy to modify; there have been numerous versions of the dashboard alone.

The Samara was cheap. All pieces of equipment in the simulation environment were intended for reuse. They are in fact still used in other research projects. Only the chassis and the PlayStation controls belong solely to the Samara. In the Steering Wheel project, the flimsy wreck of a Russian-made car with its tangle of wires, tape, and cardboard also functioned as a symbol of focusing on the essential.

Everything around me is technology—computer, projector, speakers, telephone exchange, cables, displays, and air-conditioning equipment. *I am sitting in a dim room, inside a metal chassis that was once a car. In principle, the simulation is ready, but all the task paths have to be tested and the probable error scenarios must be logged. It was already late when I started, but I have to be methodical and thorough. I start up the simulation time after time, always recording the adjustments. The simulation incorporates an audio system. Every time it starts up I hear its CD player repeating how we have the right to be carefree, and how easy money and life make you laugh. There are still things to be checked out.*

Acknowledgments

Thanks to Ville Haaramo, Topi Kaaresoja, Janne Kouri, Harri Wikberg, and Jarkko Ylikoski.

References

1. H. T. Zwahlen, C. C. Adams, and D. P. DeBald, "Safety Aspects of CRT Touch Panel Controls on Automobiles," in *Second International Conference on Vision in Vehicles*, A. G. Gale, M. H. Freeman, C. M. Haslegrave, P. Smith, and S. P. Taylor, eds. Nottingham, U.K., Sept. 14–17, 1987. Amsterdam: North-Holland, 1988, pp. 335–344.
2. M. Mollenhauer, M. Hulse, T. Dingus, S. Jahns, and C. Charney, "Design Aids and Human Factors Guidelines for ATIS Displays," *Ergonomics and Safety of Intelligent Driver Interfaces*, Y. Noy, ed. Mahwah, N.J.: Lawrence Erlbaum, 1997, pp. 23–61.

CHAPTER 10

Dawn on the Wireless Multimedia Highway

Since the early 1990s we have witnessed fundamental technology changes in telecommunications, which created growing usability challenges for mobile terminals and wireless services. The introduction of GSM had moved a considerable portion of voice telephony from fixed landline to cellular networks. It also enabled the wireless transfer of text through short message service (SMS). Further GSM evolution provided faster access to various applications through high-speed circuit-switched data (HSCSD) and General Packet Radio Services (GPRS). The Internet allowed easy and inexpensive access to a huge amount of information independent of its—or the user's—location.

The mid-1990s saw growing interest in putting mobile data and Internet services together. Downloading information through an inherently narrowband cellular channel, however, was a challenge that had not been solved in the mid-1990s; it had not even been investigated on a large scale. When the call for projects in the European ACTS Framework Program went out, we decided to tackle the problem. This gave us an opportunity to collect all the value chain players into the consortium to implement new multimedia services and try them out in a genuine second-generation cellular network. The trial evaluation was to be done from the viewpoints of all players. These presented potential benefits in understanding the customer needs and requirements.[1]

The Mobile Media and Entertainment Services (MOMENTS) project was established in 1995 at an early evolutionary phase of mobile multimedia service development, and ran through 1998. The project was started even before HSCSD emerged in a chain of digital wireless information transfer technology steps through GSM and SMS to HSCSD, GPRS, Wireless Application Protocol (WAP) and onward into future third-generation services with Wireless Code-Division Multiple Access (WCDMA) and all-IP (Internet Protocol)-based networks. MOMENTS was actually one of the first end-user field studies on third-generation (3G) multimedia services and contents using enhanced second-generation (2G) technologies: prototype solutions that were based on GSM.

Besides being one of the earliest attempts at mobile internet, MOMENTS is also an example of the methodological challenges involved in developing and evaluating new kinds of information service solutions. These solutions and their evaluation are multidimensional problems, as we'll see. The technology itself is constructed out of several layers, namely, mobile terminal, network and mobile service content production. Services are provided through a long value chain involving a number of stakeholders, and are consumed by disparate users in varying contexts. Further on, we found it nearly impossible to understand consumer responses to our solutions without comparing them to consumers' prior expectations. And since these services are so novel, consumers' attitudes toward them are subject to change as they learn and adapt. The evaluation criteria for these solutions had to be established, but it was by no means obvious what they were. Indeed, MOMENTS tells a story about working with only partially defined innovations where the problem, the solution, and the evaluation criteria are all intertwined and require fresh perspective.

The MOMENTS Project

MOMENTS (Fig. 10.1) was launched as a joint effort between all stakeholders of the mobile multimedia value chain. Our partners included content and service providers, telecommunication operators, network and terminal manufacturers, financial institutions, and multimedia technology research centers. The project objective was to demonstrate the technical feasibility and business viability of a wireless channel for the distribution of advanced multimedia products. Although the project

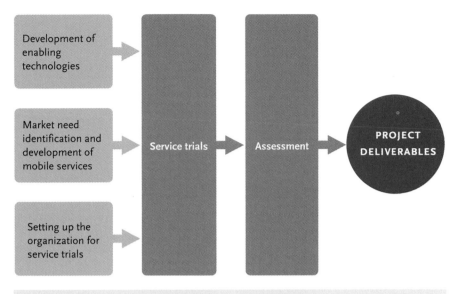

Figure 10.1 MOMENTS project overview.

focused heavily on the side of technology development, usability aspects were also studied.

The end-user trials were conducted in parallel in three European countries: Italy, Germany, and the United Kingdom. The tests were set up in much the same way that commercial services were operated—through existing cellular channels enhanced to enable multimedia service provisioning. The channels were based on mobile operators' GSM networks with several improvements that increased the speed and reliability of multimedia delivery. These trials allowed pretty realistic field assessment of wireless multimedia services and the verification of identified business opportunities.

The Nokia team took on the role of project coordinator and provided most of the technology development resources. Operator and content provider partners contributed to technology selection based on their business needs. They were also responsible for developing services, providing up-to-date multimedia files and data contents for services, and running the service trials.

The first project phases from 1995 to 1997 developed the enabling wireless technologies for mobile multimedia services (Fig. 10.2). Technology and service development turned on three main tasks:

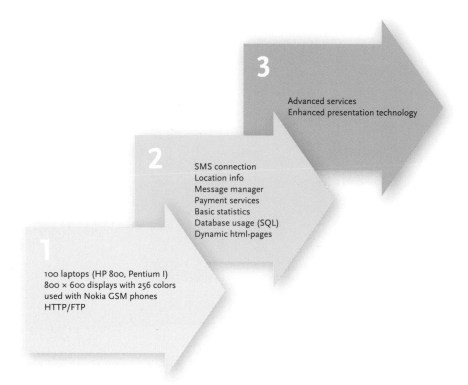

Figure 10.2 Development phases of wireless multimedia enabling technology.

- Constructing mobile terminals capable of receiving and displaying mobile multimedia
- Enhancing wireless multimedia transfer technologies to allow fast and reliable download
- Creating services providing multimedia content especially attractive to mobile users

How We Went about It

In those days there were no digital cell phones capable of presenting multimedia contents. Consequently, mobile client prototypes were constructed using Hewlett-Packard 800 series laptops (Pentium 100 MHz, VGA 256 colors displays; see Fig. 10.3) for content presentation and manipulation, and Nokia GSM phones supplied information transfer. This

Figure 10.3 MOMENTS mobile client prototype: Hewlett-Packard 800 series laptop with Pentium 100-MHz processor and VGA 256 colors display; and a Nokia 8110 GSM phone.

prototype was viable for use in a semimobile usage context, such as at home, in a hotel room, or at an airport. It was possible and relatively convenient to carry the set around, but establishing a connection took awhile, and using the services with the laptop interface required the user to remain stationary. Nokia 9000 Communicator, which was the first product that would have allowed real mobile use of mobile services, was launched just as the project was ending.

Several new functions had to be developed for the client terminal to enable consumption of multimedia services. These included

- Multimedia presentation optimization for low bandwidth[2,3,5]
- Terminal positioning capability based on GSM cell identification

- Direct local-area network (LAN) access to the mobile switching network for optimal server connectivity and for the improved reliability of the connection[4]

Data transfer through the cellular channel was also enhanced by several means:

- Automatic reestablishment of connections after breakdowns
- Session recovery
- Traffic prioritization
- Optimized client start-up
- The use of data compression technologies

Off-line and real-time encoding and decoding tools were developed for selected audio and video compression formats.[5] Development of graphical information presentation tools was concentrated on the combined presentation of 2D/3D vector graphics, animation, and on the format conversion, adaptation, and content creation.[2] Scenes with graphical content overlays could be automatically created on the fly for mobile services.

Our selection of services to be included in the trials was based on market research carried out in the participating countries, and on the enabling technologies available in each study phase. The following types of mobile multimedia services were provided to end users:

- Unlimited access to public Internet services
- Recommended Internet services selected by the operators for mobile users from the public Internet
- Advanced mobile services developed specifically for the MOMENTS trials[6]

The content providers in each trial country designed the advanced mobile services (see Fig. 10.4 for a partial list) using the enhanced technologies mentioned above. Premium business services were expected to be the most popular ones at the first penetration phases of the mobile Internet. Thus, service selection criteria were very much on the side of utility, with some minor exceptions.

CATEGORY	SERVICE
NEWS	Politics
	Economics
	Local, domestic, and international news
	Sports
	Culture news
CITY INFO	**Street search and traffic update**
	Location identification
	Local transportation
	Hotels
	Restaurants, bars, and discotheques
	Entertainment sites, movies, theaters, concerts, and arts
TRAVEL	**Highways information**
	Best route selection
	Flight information
FINANCE	National and international main indexes
	Stocks, Futures, and Currencies
	SMS alert triggered by indexes and stock value changes
HIGHWAYS SERVICE	Maps
	Traffic update
	Location identification
	Targeted SMS information
WEATHER	Daily forecasts: National and Europe
	Animation showing moving weather fronts
	Local weather evolution
ENTERTAINMENT	TV programs scheduled on public and private networks

Figure 10.4 Wireless multimedia services used in MOMENTS field trials. Advanced services developed especially for mobile use are denoted by red text.

Let's take a closer look at an example of an advanced service developed within MOMENTS.

Milan-Area Traffic Service

An interactive map of Milan (see sample screen in Fig. 10.5) became the basis for all location services related to the city. The map was preinstalled into the client computer, and only the data updates were made over the air while mobile. To use the service, a customer accessed the service homepage, opened a street map encompassing the whole metropolis, and could then zoom in and out by clicking on icons, or pan in all directions by clicking on the edges of the map. Panning was particularly useful while following the path that a particular street cut through the city. Character shortcuts and arrow keys were provided as alternative ways to navigate.

Location identification is a feature telling customers their present location on the basis of the mobile network cell's identification information. A query is launched when the customer clicks on a "Where am I?" icon. Our system generated a reply in 6 seconds including date and time, geographic coordinates of the center of the cell, and an approximate radius of coverage of the cell. The map was then redrawn to indicate the user's approximate physical location. This information was presented by displaying a circle with the given radius centered on geographic coordinates of the cell's base station. The radius indication was seen as useful, because the radius of the cell varied from city center to peripheral areas. At the same time the date and time of the actual location was displayed in a window, as the user may have been moving. The accuracy of location identification was not precise enough for street navigation, but the approximate location helped a driver interpret traffic information presented on the interactive map.

Traffic updates were recalled from a database on the server. The database itself was continuously updated by a specialized traffic information service provider. Topical traffic conditions were transferred in a very compacted format to allow the fastest possible download time to the mobile client. A typical traffic update download was 5 kilobytes (kbytes) with a download time under 8 seconds, including map redraw time. This

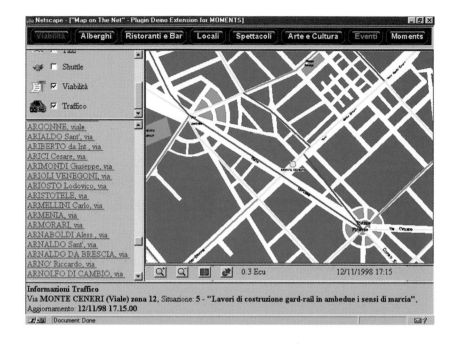

Figure 10.5 Milan-area traffic service sample screens.

extremely compacted traffic data was available for about 3000 streets around the Milan area. Each street's traffic status was indicated by a different color—green for light, yellow for medium, and red for heavy traffic. In addition, the map showed specific traffic events. By clicking related icons, users could invoke detailed text notifications to be displayed on the bottom frame of the screen. Information depth was increased with impediments such as street construction works in progress, traffic accidents, fog, and heavy rain. These downloads included about 150 characters of text about the location of the event, a short description, and the exact time of last update, and download time was no more than 3 seconds.

Designing the Tests

Service trials were run from March to July 1998. The setup was arduous, including the installation of client/server systems, the selection and training of end users, definition and implementation of the service provisioning chain, issuing equipment to end users, and planning questionnaires, statistical methods, and help desk services. Several company-internal pretrials were therefore conducted to fine-tune the process.

The trials were organized as a series of test periods about two weeks long. The total number of participating end users was 70. The amount was to be maximized for sufficient statistical reliability but practically limited by manageable amount of persons during trials. This number was seen sufficient for our purpose. Since the number of users was limited—and segmentation by country was already a given—we agreed to focus on one user profile instead of delving into user segmentation according to gender, age, profession, and other criteria. In line with the business context of the services to be tested, the following target profile of an end-user was defined:

- Male; in the product development units of the participating companies, the male gender seemed to be dominant in test user selection.
- Midrange age, mostly between 20 and 40 years.
- High level of mobility.
- High level of education.
- High demand for communication.

Internet experience and GSM usage were also part of the selection criteria. Thus, the whole end-user view was very business-focused, reflecting the developers' expectations of the likely order of adoption for mobile multimedia services.

Typically the first series of precommercial equipment and services have not gone through field testing. The instability is likely due to varying and demanding mobile environment conditions. Thus the user-friendly test approach was chosen to start with. The users were selected from those having experience on Internet, mobile phones, and services usage. Thus they were more ready to cope with the technically demanding situations (e.g., contacting help desk services) instead of quitting the trial fully in despair. In the first phases of the trials, the subjects were personnel of the participating mobile operators. In later phases company-external users were included as test subjects.

We applied user-centered design approaches to improve the system quality as experienced by the end user. A number of design methods were chosen to support the development in various design, evaluation, and user support tasks. Focus groups discussions were conducted in early development phases to provide an indication of users' service preferences. Then around 40 to 50 rough service ideas were generated in several brainstorming sessions. Storyboards were used for more detailed service concept design after the core ideas had been identified. Content providers developed use cases to help us understand the users' tasks on a step-by-step level. Low-fidelity prototypes were constructed to evaluate early UI versions for service access, and usability tests were carried out to assess the solutions.

In addition to designing user interface solutions, usability evaluation criteria were developed to suit the specific needs of the project. Examples of usability criteria were user interface self-descriptiveness and feedback, controllability, and flexibility; efficiency, safety (mental, physical, and property), error-freeness, and error recovery; clear exits; and functional consistency. However, the field tests in MOMENTS were not intended primarily for optimizing the usability of a new device, but for assessing the overall acceptability of novel technologies and new services provided through them.

The user's perception of the quality and acceptability of a mobile service is always influenced by a multitude of factors. First, there are the mobile client, the applications in it, the wireless connection to a server, the service presentation, and the content. Also, there are the real performance of the system and the expectations the users have for it.

Expectations about brand new technologies are influenced by another set of issues. Personal experience with similar or comparable technologies forms an obvious base of reference. (In the case of wireless services, the user's experiences with the wired Internet was a natural and relevant comparison.) Personal needs between individuals vary—whether professional or private—setting up different criteria for acceptability. The type of service itself has some effect on how it will be evaluated. Information services, for example, are assessed on a different basis from entertainment services. Finally, there are market driven expectations about services, which are conditioned by their pricing and the reputation of the provider. We needed a conceptual model to link all these factors affecting service quality (see Fig. 10.6).

At this stage we had no way of knowing which quality dimensions would be the most influential, so the project needed to address a wide scope of customer experiences. The following list of evaluation criteria provides some idea of the complexity of assessing mobile service quality.

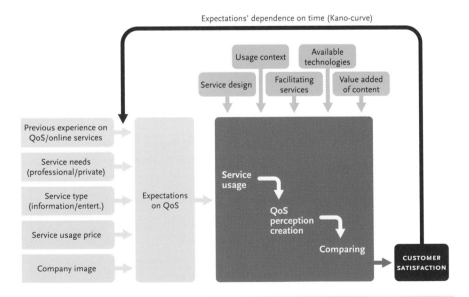

Figure 10.6 Quality evaluation model of MOMENTS services.[9]

Our challenge was made all the more difficult because we didn't know whether quality criteria would vary across different services and test sites.

- Usefulness of the service to the consumer, and the concomitant ability to tailor the service to personal needs and link the service to the user's business and private objectives. We ask "Is the content what you needed? Does the service provide something just for you?"
- The amount, versatility, and accuracy of content provided in a given service can be judged only according to what service it is. In a location service, for example, accuracy may be a critical quality criterion. The timeliness of the information provided by the service is another potentially important dimension related to accuracy.
- Response times in interactions are contingent on connection establishment, information download, and release time. Most people have expectations for the speed of service according to their experiences with the wired Internet. Mobility can be an acceptable excuse for somewhat slower respond times, but there is no doubt a limit that cannot be exceeded. We were also aware that the perception of acceptable delay varies with the type of interaction; for example, real-time game application does not allow the same kind of delay as the email services.
- Problems in connection reliability can undermine an otherwise perfect service. A mobile service in business context has to be perceived as reliable. If not, users will cleave to conventional information sources. Usage patterns that do develop may take the service in a completely unexpected direction.
- The quality of content presentation in different media formats such as audio, video, animation, images, text, and graphics will probably be compared to expectations set by other known media. In addition to the user's previous media experiences, synchronization of different media formats affects the overall quality perception.
- The way the user interface is designed can influence the user's perception of security and privacy in a service, but the public image of the company providing that service also plays a role in determining that person's level of trust in the service.
- The mobility of the terminal in practical usage situations is an important assessment criterion, but differences between the test set and late-

model compact terminals can cause biases that need to be acknowledged.

• How easy it is to find required information and how appropriate the access methods are—these factors cut close to the core of basic usability.

The evaluation pack for MOMENTS was designed to give us a good idea about the acceptability of services with reference to these criteria in broad strokes, rather than going into deep detail with any individual issue. Usability data in all three trial countries—Italy, Germany, and the United Kingdom—were gathered using pretrial and posttrial questionnaires, face-to-face interviews, and questionnaires filled out by help desk personnel.[7] The effective use of those approaches called for a commitment from all the partners in the project. Since MOMENTS was a large international project with multiple partners at different sites, the responsibility for usability assessment was spread over several locations and development teams. Nokia prepared the usability and quality evaluation guidelines for the project, and operator partners carried out the fieldwork. The mobile operators applied the guidelines to suit local conditions (e.g., operator-specific requirements). The original objective was to gather comparable results from all participating countries, but it turned out that the evaluation approaches had to be localized just like the services themselves.

Distinguishable Services and Critical Mass

In the field trials the overall rating of the quality of mobile service provisioning was positive for most end users in all countries. The highest quality ratings were given to such quality dimensions as reliability of connection, clarity of status information during use, and ease of use of the services. Most users found the services and equipment to be easy to use and attractive, and to furnish sufficient navigation methods. The end users appreciated in particular automatic reestablishment of connection, HTML page transfer, SMS services, and clearness and richness of information. Their criticisms concerned the slowness of loading the program and the need for more compact equipment. Expectations for mobile multimedia technologies seemed to be independent of country.

Expectations for service content, however, depended more on local culture and on the way the trial services were implemented in different countries. Three key criteria for successful service content provisioning were isolated:

- Time-saving capability
- Distinguishable from existing media channels
- Presenting the user with a critical mass of service options

The mobile user, a business user at least, experiences the service as beneficial if it is possible to save time with it. Entertainment services, on the other hand, are beneficial if they help the user *spend* time during idle periods. Service airtime and subscription expenses aside, busy users will regard their efforts to install and maintain the service software, and to download content from it, as additional costs of use. Any possible time savings the service yields may be negated if maintaining the systems starts to take too long.

Among the services tested, city information, traffic information, and short message notification of traffic conditions were regarded as very attractive. Traffic information offered data that was not otherwise accessible. Weather and news services, on the other hand, were not considered attractive. Services in which subjects exhibited less interest were not sufficiently distinguishable from existing media channels, namely, TV, newspapers, and wired Internet services. Competition between information channels gives the end user several choices besides the mobile Internet.

Consistent with those findings, location-based services have added value with which to attract users in comparison with the general-purpose Internet services. In Italy, where the Milan traffic service was offered, users saw services as more mobile-oriented whereas in Germany and the United Kingdom only general travel information was provided (see comparison in Fig. 10.7). Thus, overall satisfaction in Italy measured higher on all scales. Traffic service is also an example of a time-saving service.

To sum up, mobile access cannot in itself be assumed to be a remarkable benefit for every category of services. The benefits were perceived only when content was specially tailored to fulfill needs in mobile usage situations.

	UK	GERMANY	ITALY
Speed of starting system	●	○	○
Speed of downloading services	●	○	○
Reliability of connection	○	○	○
Clarity of status information during use	○	○	○
Service ease of use	○	○	●

Figure 10.7 The ratings of services in different countries reflect the differences in service implementation. Overall rating on developed mobile services in Italy was clearly more positive than that of a selection of public services provided in the United Kingdom.

Another finding with respect to the acceptability of mobile multimedia services is that it seems to hinge on the availability of a sufficient number of advanced mobile services. The availability of new technical solutions is not sufficient to attract consumers unless there is enough drive on the service provisioning side as well. Once providers exceed critical mass in the variety of services offered, the users are able to find the ones that match with their personal preferences, needs, and expectations. They begin to be willing to access mobile services repeatedly, which leads to behavioral learning and to the adoption of those services. Of course, a personalized service portfolio is possible only if the selection is big enough and the user has the freedom to choose. Note that this means that a single user is not supposed to continuously consume a large number of different services; range is necessary for offering the appropriate selection.

Efficiency of the Trials

Field trials achieved a semimobile usage context, which was a much more realistic situation than in laboratory studies. The users alone were responsible for activating services, whereas laboratory studies guide the subject through the process, and there are fewer opportunities for user-initiated decisions concerning use.

The business case for wireless multimedia services was less convincing for operators during MOMENTS than it is currently. Doubts about investing in mobile multimedia could have affected both motivation and resource financing as the project advanced into more demanding phases. Under these circumstances, the field trials gave the operators a chance to get deeply involved in testing and let them develop some competence with the technology, the establishment, administration, and analysis of the trials.

We needed to judge the costs of the project for each partner. The efforts in person-months were calculated over 3 years after the actual project and follow-up reports in order to determine project efficiency. These numbers included all phases from the initial organizing of trials, through the fieldwork and time spent evaluating study results. The division of effort in the service trial project phases was 35 percent for the organization of field trials, 33 percent for the operation of field trials, and 32 percent for the evaluation. This was a good result; if the investment in organizing and running trials is inadequate, the validity of results decreases because of insufficiency of data. Increasing the evaluation efforts later on cannot improve results where usage activity has been too slight.

In large trial projects it's hard to estimate costs and efforts needed in order to reach all original objectives. In this project the funding was shared between partners, and in fact two new partners were invited into the project at the beginning of the second project year. The lesson learned for similar projects is that flexibility in replanning is essential for the entire project duration.

Conclusions

MOMENTS reached its objective to test the technical feasibility of wireless multimedia in an early phase. As a result of this project, commercial information technologies were applied to wireless multimedia services, new audiovisual enabling technologies were developed, a wireless payment system was demonstrated, and the first multislot GSM was deployed. The Internet was experienced first as a competitive but also as a synergistic opportunity for mobile services creation. Partners were deeply involved in the trialing from beginning to end. Postproject cooperation with related projects and across the mobile multimedia technology clusters improved the new business partnerships and helped focus business strategies for the future. We believe that these trial experiences were a useful advisory for planning trials for the development of mobile 3G multimedia.

In the planning phase of 1995, the vision of wireless services and technology solutions was not clear, and thus the preliminary character of these trials is evident in retrospect. Instead of marshalling a well-defined set of services, technologies, and study assumptions on end-user behavior, we took an approach to the studies that was more innovative than systematic. This led to high ramp-up on participant competence improvement, which was utilized well in follow-up projects. Thus the value of the project can be stated not only in terms of itself but also as a part in series of business development actions.

Acknowledgments

The authors express their gratitude to EU, ACTS Commission, and to all project participants in Nokia Oyj, E-Plus Mobilfunk GmbH, Orange Personal Communications Services Ltd., Zentrum fur Graphische Datenverarbeitung eV, Reuters AG, Gemplus International SA, Bertelsmann AG, Omnitel Pronto Italia SpA, Citicorp Kartenservice GmbH, DataNord Multimedia SrL, and all the mobile citizens generously contributing to the trial operations.

References

1. J. Pekkarinen and J. Salo, "Wireless Multimedia Research, Balancing the Efforts in Service Trials," *IST Mobile Summit 2000*, Oct. 2000.

2. N. Gerfelder, H. Jung, L. M. Santos, and C. Belz, "Challenges to Deliver 2D/3D Content for Multimedia Applications in Mobile Environments," *ACTS Mobile Summit '98*, June 8–10, 1998.

3. M. Karczewicz, L. Oktem, and J. Nieweglowski, "Advanced Video Coding Scheme for Low Bit Rate Applications," *ACTS Mobile Communications Summit*, Nov. 27–29, 1996.

4. T. Lindgren and J. Lahti, "Challenges in Delivering Multimedia Services over the Cellular Channel—Requirements for and Implementation of Robust Link," *ACTS Mobile Summit '98,* June 8–10, 1998.

5. M. Luomi, "Audio and Video in the MOMENTS Project," *ACTS Mobile Summit '97,* Sept. 1997.

6. G. Melpignano, "UMTS Services Applications," *IEEE Commun. Mag.* (Special Issue on UMTS) (Jan. 1998).

7. K. Väänänen-Vainio-Mattila, "Aiming at MOMENTS Result Validity through a Complete Set of Research Methods," *ACTS Mobile Summit '98,* June 8–10, 1998.

8. X. Yu, "Mobile Multimedia Services Based on GSM HSCSD," *ACTS Mobile Summit '98,* June 8–10, 1998.

9. N. Kano et al., "Attractive Q. vs. Must Be Q," *Hinshitsu,* **14**(2): 39–48 (1984).

John Rieman

CHAPTER 11

Just-In-Time Usability Engineering

With its leadership position in the market, Nokia has the luxury to influence many of the standards, real and de facto, that define today's wireless user experience. This position has been attained with the aid of skillful designers, innovative contextual design methods, close ties with leading researchers, and a fortunate location in a country that is unusually receptive to new technology and lifestyles.

But this influence has limits. Key interactive technologies are surfacing worldwide, many of them outside Nokia's sphere of influence, and some of them even outside the company's knowledge. Some are user-neutral and many are user-hostile, and as usability engineers, our natural response is to decry the lack of user participation in their definition. "Why weren't we asked?" We shake our heads. "Why didn't they consider users' real needs?" "Who let the bloody technologists define this monstrosity?"

If you are in the business, you know the examples: WAP and its associated bundle of acronymic standards (WIM, GPRS, WMI, etc.). Of course m-commerce will simplify our future lives, but why is it so incredibly hard to get it running today? The service providers blame the handsets, the handset designers blame the service providers, and the user is in the middle. If we had to buy groceries this way, we'd all starve before we got the accounts set up.

First, consider SyncML. The need to synchronize data across several devices is obvious, and Nokia is a key member of the group defining this critical standard. It works, but these are the early days, and the underlying technology still has something of a "camel is a horse by committee" flavor. So many options, so much flexibility—how do we incorporate that into an elegant solution for the average user, who just wants her one-and-only PC phonebook to show up on her one-and-only phone?

And third-generation cellular systems with "rich-call" technology—it's hard enough to drive a car while talking, but now we can simultaneously set up a conference call, check our flight time using the browser, and monitor our position on GPS. Never mind the user interface; we'll need three people just to *think* about all those things at once.

But while the nay-sayers and Luddites shake their heads and mutter, these technologies are opportunities ripe for the taking. Many of today's "unusable" applications will define tomorrow's market and user experience. This is no small issue. In the short history of the digital world, the examples of "ugly duckling" technologies that transformed into swans are legion.

Consider email—in its earliest incarnations, it was just a way to pass a note from the late-night computer operator to the early-morning computer operator: "Tape drive 3 dead—joe." A mail message couldn't move beyond a single, one-user machine. That was in the days when programmers wrote code on paper pads and handed it to keypunch operators, when you carefully penned a diagonal stripe across the top of your card deck in case you dropped it, and the man (person) in the street thought of computers as walls of flashing lights on "Twilight Zone" episodes.

Contextual design of email for the office environment? Hardly. But it wasn't needed, because email grew like a living thing, as affordable machines, timesharing, local-area networks, and ultimately the Internet provided new opportunities to be exploited by late-night technologists.

Or think of spreadsheets. Did the initial requirements include what-if planning? Database facilities? Interactive three-dimensional graphs? The original designers could hardly have dreamed of such power.

And the Internet infrastructure—nuclear strike–hardened, designed by computer scientists to exchange software and raw data, a technological tour de force. But e-commerce? Free HotMail? Buffy fan fiction? Not on *our* tax dollars!

There were microprocessors, global-positioning systems, code-division multiple-access (CDMA) cellular technology (you couldn't have done the encoding computations in real time on a 1950s mainframe), and a hundred other examples. Sure, there were futurists, science-fiction writers, and other visionaries who imagined how it might all turn out. But even they were usually wrong. As recently as the early 1970s, Kubrik and Clarke's *Space Odyssey* seemed so plausible.

Of course, as each of these technologies coalesced, our formal design techniques were able to encompass and reform them. But in their dawning days, it was the late-night technologists who made things happen, for their own strange reasons, in their own strange ways. They are still with us, pushing the envelope of the wireless world just as they pushed the envelope of computing. Their innovations are the new bricks and mortar with which we will build the wireless information society. We may shudder or marvel at the complexity and idiosyncrasies, but to support our users and the company's business, we must work within these technologies even as we try to change them.

Just-in-time usability engineering lets us do exactly that.

Just-in-Time in the Factory

Step back in time to 1950. Japan, devastated by war but eager to reenter global markets, examines its factories and management practices. At Toyota, Taiichi Ohno finds that the company's factories are only fractionally as efficient as their American competitors. Ohno and Dr. Shigeo Shingo, inspired by American efficiency experts such as W. E. Deming and Frank and Lillian Gilbreth, examine the entire factory process and propose a radical restructuring. The goal is to reduce waste, improve quality, and ultimately return Toyota to profitability.[1] The new approach to the factory will be known by its central tenet, "just in time" inventory management.

But it includes other key factors as well. It incorporates an unprecedented emphasis on quality, and a fundamental respect for the humanity of the worker. It is instrumental in bringing Toyota and Japan back to the world stage as an industrial superpower. By the 1980s this was the pattern that American industry hoped to emulate or outdo.

Just-in-time inventory management and its characteristic *kanban* control cards are at the center of it all.[2] As the term suggests, *just in time* describes a production system that does not rely on a large inventory of components stored at each factory or in central warehouses. Instead, each production station maintains a small store of only those items it needs for orders currently being processed. When the inventory becomes low, the station places an order from its supplier, and the supplier is expected to fulfill the order almost immediately. Production flow is controlled locally, using the kanban inventory-pull cards, instead of being planned and monitored by a central system.

The most obvious advantage of the system is that the cost and potential waste of maintaining a large inventory is reduced. The company buys only what it needs from its suppliers, and it makes the purchases just in time to use them. In a factory setting, supplies may be requested several times a day. The suppliers themselves may be production units within the factory, or they may be external to the company. Capacity of suppliers and the main factory is reduced to exactly what is required.

Beyond inventory cost control, an equally important goal of the system is to provide significant improvements in quality.[2] Toyota strives for zero-defect final products, since any defective product reflects on the quality of the entire brand. The small lots from suppliers allow each item to be inspected at the production station before it is used. Defects are returned immediately to the suppliers, representing a quantum improvement over the traditional approach in which thousands of items may be purchased and warehoused with only spot checks for quality. Because defects are returned immediately, suppliers can quickly correct any error in their processes. This also encourages a continual improvement of process and quality, referred to as *kaizen*.[2]

To further reduce waste, workers are expected to be skilled in many tasks. Not only do they perform many of the set-up and clean-up activities associated with their workstations, but they are able to step in and help with coworkers' tasks when demand requires. This allows skilled labor to be supplied "just in time" as well.[3,4]

The multiple skills of workers, the team responsibility for the quality of output, and the dependence and trust in suppliers all recognize the importance of the workforce. Workers are expected to have a profound understanding of their tasks, and their suggestions are the key to making *kaizen,* the continual improvement of the process, possible. The worker's proactive participation is essential to the just-in-time factory.[5,6]

By the 1980s, the Toyota approach was so successful that it was imitated by many American and European factories. Corporations using just-in-time or "zero inventory production" systems included Hewlett-Packard, General Electric, Tektronix, and many others. Related practices include quality functional deployment (QFD), total quality management (TQM), and quality control circles (QCCs). Some of these peaked as fads and were later discounted, but the core directions remain very much alive.

Although just-in-time processes were designed for the factory, many of the ideas are also applicable to service industries. A similar focus on quality, efficiency, and zero-defect production has been the key to success for private mail services such as Federal Express. Even accounting firms have found it applicable. And as we shall see, it is well suited to user interface design within the software factory model used by Nokia.

The Problem with Traditional Usability Engineering

The traditional American approach to usability engineering was first defined by John Gould and Clayton Lewis at IBM almost 20 years ago: Investigate users and their tasks. Design and prototype. Evaluate and measure. And always: iterate the design and test again.[7]

A generation of researchers have suggested refinements and minor shifts in focus. At Virginia Tech in Blacksburg, John Carroll prescribed scenarios;[8] at Bellcore, Thomas Landauer championed metrics and focused iteration;[9] Clayton Lewis and John Rieman at the University of Colorado emphasized tasks;[10] and Jakob Nielsen (Danish in heritage, but taking an American approach at Bellcore and Sun) compromised with discount testing while retaining the overall direction.[11]

There is also a European version of the tradition, more humanistic, less clinical. The user is regarded not as an object to examine but as a participant in the design process, a partner in the contextual inquiry, a collaborator in engineering a system to be satisfying as well as effective.[12] But it is only a shift in emphasis. Users in American labs are human as well, and Americans have added significant value to the "European" approach.[13]

Whether American or European, both traditions work best in a world that is increasingly rare, a world where one can, as the king advised Alice, "Begin at the beginning, and go on till you come to the end; then stop." The signature publications of these traditions reflect this viewpoint, describing showpiece or exceptional situations where end users are surprisingly available and schedules are surprisingly tolerant. IBM's voice messaging at the 1984 Olympics in Los Angeles is an example,[14] as is Nynex's ambitious project to redesign long-distance operator workstations,[15] and the Utopia project in Scandinavia, creating early tools for computer-aided layout and typography.[16]

Whenever they can be applied, these careful, long-term, task- and user-centered approaches will continue to have value within Nokia and the telecommunications industry. Many of the chapters in this book describe ongoing work in exactly that vein. But in the world of consumer products, we increasingly find ourselves pushing the envelope of user behavior, providing support for future tasks that today's users haven't even imagined. In such areas, the usability engineering must run concurrently with the underlying technology—and often behind it. The work can happen just in time at best, although calling it "just in time" risks the criticism sometimes levied at Java just-in-time compilers—the work may happen just *after* it would be most useful.

This is the world today, and like Japanese industry in 1950, we must compete in the world as we find it.

Just-in-Time Usability Engineering

Today's technology may change too rapidly for traditional usability engineering. The designer can't identify users and their tasks before design, because the technology being designed will create its own tasks. In many cases, the application is strongly defined by standards that showcase the technology, leaving the designer to imagine what it might be good for.

Specialist application teams and just-in-time usability allow Nokia to produce usable systems even under these constraints. And although Japanese factory management techniques were not the inspiration for these design processes, the similarities are striking. The clearest parallel is to just-in-time supply requests in the factory. In the same way, Nokia's usability experts must respond quickly to requests for information or evaluation. There is seldom a backlog of usability studies to draw on, because most situations were never planned or imagined.

> *A designer calls the team's usability expert: "The wireless carrier's messaging standard has just changed," he says. "We can't send as many characters as we thought, but we can still receive them . . . is there a usability problem?"*
>
> *Or, "We're using a nonstandard display on this product, and the mechanical design can't be changed . . . now we need a readable font."*
>
> *Or, "We've designed this new browser add-in to meet the common standard . . . can you help us figure out what users might do with it?"*

This is not the traditional, carefully planned, cradle-to-grave iterative design approach. This is reactive engineering, where the rational assumptions may be toppled at any moment like a line of dominos. And to allocate an army of researchers to cover the range of possibilities with prospective studies would be wasteful, indeed.

To handle these sudden, unplanned requests, the usability experts—I'll call them the "platform usability group" to distinguish them from UI researchers—must be generalists. Like the Japanese factory workers, they must be ready to pick up any task, with help from their colleagues and research groups in sites scattered across the world, and provide an answer, or even a firm "don't know," at the shortest notice.

> *"This is the new hardware proposal . . . we have to make a go/no-go decision now. Are the keys OK . . . or does it need testing?"*
>
> *Or, "The wireless operator wants this new message during roaming calls. It has to be in the software release for next week. Are there usability issues here?"*

The hardware proposal is an example that exemplifies the just-in-time approach. The industrial design for the product had been completed and evaluated months earlier, but changes in the predicted market and the schedules of other products demanded a sudden redesign in a smaller form factor. The redesign was completed in record time. Coincidentally, the new wax model was presented to management a few days after the first working prototypes of another product were distributed to internal users. Users had begun to report that the key layout on the working prototype encouraged a fairly serious type of error, and a redesign was being considered. Managers now were worried that the design represented by the wax model, which had a key layout similar to that of the prototype, would produce similar usability problems.

Usability testing could shed light on the question, and every usability expert has the expertise to run the tests. But there was no way to test for key press errors on the wax model, and a decision had to be made immediately. Usability experts turned to their knowledge of human-factors techniques and performed *link-analyses,* evaluating the hardware with reference to usage patterns of keys in the standard Nokia UI. They concluded that even if errors did occur, the risk associated with them was low—the user could easily recover. The industrial design group was able to make minor changes to reduce the chance of error even further, and the design was approved.

It is no easy task to give immediate evaluations of products that stretch the boundaries of technology. The educational backgrounds of the platform usability group reflect this broad challenge. Most have at least a master's degree; several have additional education or PhDs. In addition to the obvious disciplines of human–computer interaction and various behavioral sciences, the group includes students of economics, anthropology, and computer science. Some are programmers, some have studied statistical techniques, and some have managerial strengths. Several are trained in hardware ergonomics. All must be familiar with Nokia's UI styles and basic features. All interact regularly over the Net (Internet) and in person, despite their global distances and time differences.

A usability engineer in Nokia's San Diego site comes in at 5 a.m. several days a week, to join email and phone conversations with colleagues in Europe. "There's also less traffic on the freeway," he notes.

Responsibility for quality is a third point where the Japanese and the Nokia approaches agree. Just as factory teams are expected to maintain defect-free quality in their output, each Nokia application team is ultimately responsible for the usability and overall quality of its own work (an approach pioneered by Gene Lynch and his colleagues at Tektronix.[17] The key team members are the UI designer and the software engineer. There is a localization specialist to define texts, a platform usability expert, and test engineers, and there may be graphics and sound support as well. For complicated applications, the user guide author may even join the team.

The team's multifaceted background makes it ideally suited to perform various forms of structured usability inspections, originally developed through research into user interface design processes in the 1990s.[18] Nielsen's formalized version of the traditional "heuristic analysis" is probably the best known of these. Heuristic analysis, performed in many fields, involves examining a system for violations of clearly stated guidelines, or heuristics. For UI engineering, Nielsen suggests a small set of widely applicable heuristics, such as "Prevent errors." His research also shows the value of combining comments from several inspectors. The analysis can be especially productive when the inspectors have "double" expertise, covering usability as well as the system domain area.[11]

Walkthroughs are another inspection method arising out of existing practices, in this case the code walkthroughs of software engineers. The "cognitive walkthrough" is designed to apply theoretical work by Clayton Lewis and Peter Polson. Each walkthrough involves a step-by-step examination of the sequence of mouse clicks or similar actions required to complete a typical user task with the system under design. The inspection process uses specific questions and criteria to help the inspectors predict the mental behavior and likely errors of a novice user.[19] Randolf Bias' "pluralistic walkthroughs" are another approach, placing less emphasis on cognitive theory and more on the social dynamics of bringing critical stakeholders together in the face-to-face inspection meeting.[20]

The inspections are done by experts, without involving real users. This allows them to be completed well before code or even simulations are written. Combined with on-line inspection of design documents by other teams and usability experts, they form a first line of defense against egregious usability errors. The inspections also help focus subsequent user testing on critical issues. Often these involve texts or icons in the display. If the inspections identify the problem areas, alternative textual or graphical items can be tested using surveys. This provides a wider coverage of markets and user groups than does task-focused testing in a usability lab. Sometimes the situation is more complex, and the user's behavior depends on a mixture of background knowledge and problem-solving techniques that is difficult to predict without observing real people. In these cases, more traditional usability lab methods can be used.

> *Reviewers of an application that is patterned on its PC counterpart note that one screen has an unusual mixture of user data and default functionality. It seems to be what's needed, but will the user be confused? The inspection raises the question; we'll consider user testing as a route to the answer.*

The usability expert joins the inspections and may arrange testing, but never acts as a final authority. In yet another concordance with the Japanese model, the work is a team effort. The UI designer and the software engineer are typically the final links in the decision chain, but this reflects operational necessity as much as formal authority. Usability issues can be discussed by any member of the team, and the final design is an informal consensus.

> *A software engineer resolves a data-calls usability issue by applying the "mom" test: "If my mom wouldn't know what to do with this menu, we don't need it." The team decides that she is right.*

The involvement of the entire team, especially the interaction designer, provides a baseline of usability that makes it unnecessary to usability test and iterate every element of each new product—an impossible task with a product line on the scale of Nokia's. And even though plans and resources are allocated for testing major new designs at several stages, the focus of the testing is frequently revised to address controversies that have just surfaced.

> *A critical series of screens in a new UI style is designed to use icons as control elements. The aesthetic improvement over a purely textual interface is evident, but there are lingering questions about the usability of the icons. An already scheduled usability test is replanned to focus explicitly on the issue.*

The text-versus-icons study is a good example of the just-in-time approach. This question had come up earlier, and was even tested as part of an exploratory effort considering several new designs. The chosen design incorporated many novel features, and it was expected that several rounds of design, testing, and redesign would be required. Progress was being made, but after two rounds of testing, users still had problems in coming up to speed in the new system. The issue of the icons was raised again.

Another round of testing in several countries was already scheduled, but the test plan was modified in the days just before the test. Instead of exposing users to the latest design and watching for problem spots through the usual think-aloud protocol, two alternative designs were tested: one with icons and one using texts instead. Task times and errors were measured and analyzed statistically. The test results alone did not answer the question, but they tipped the balance to allow a decision that all parties could support without further discussion: The text-based design was selected for further development, and the iconic system was postponed until further research could make it acceptably usable.

Opportunistic actions such as revising existing test plans "on the fly" are another defining feature of just-in-time usability. The opportunistic approach is reminiscent of the opportunistic planning described by Barbara Hayes-Roth and Frederick Hayes-Roth in artificial intelligence (AI) literature.[21] It allows the small team of platform usability experts to make the most effective use of their time, rather than following a slavish routine of usability evaluation and testing for every design, no matter how trivial. It also allows the group to emphasize another goal that parallels the Japanese approach: continuous improvement.

A strictly planned route to continuous improvement would entail a lifetime cycle of design, implementation, and tracking of problems in the field. Some of this is possible, but it could take years to feed changes back into the next version of a product—and by that time, much of the technology may be obsolete. The opportunistic approach and Nokia's platform support for product development allow this cycle to be compressed, by feeding information from several overlapping product cycles into the development effort.

An implementation of instant messaging on a limited-market prototype is studied with internal users. The technology in this product is already outdated, but studying it raises flags that will help avoid problems in the revised system now under development.

The "out of the box" experience for a new product is tracked with diary studies for several months, to identify usage patterns and major difficulties. No significant changes will be made to this product on the basis of the study, but the results feed into our general understanding of design. (One finding is that mobile phone users do read their user guides when faced with a difficult task. This will help designers decide how to distribute information among the user guide, on-screen help, and other resources.)

This example points to another situation where the usability group reacted to a fortunate set of circumstances. The project was a joint effort between Nokia's usability experts and a master's student doing thesis work in a local university, allowing the usability team to increase their effective resource count temporarily exactly when it was needed. Interestingly, part of the study was patterned after similar work that had been performed with personal computer users almost a decade earlier—a study that had been the dissertation project of one of Nokia's usability experts.

The Zen of Just-in-Time Usability

There are process definitions for UI design and usability engineering within Nokia, as required by standard engineering practice. However, these processes are typically presented in terms of required documents and inspections, supporting tools, and checklists of issues—not as step-by-step procedures to follow in every case. Within the defined structure, activity sequences and even responsibilities may vary considerably. The interaction designers and software engineers share primary responsibility for producing a usable system. The role of the usability expert is to support the design work.

Figure 11.1 shows a number of points where usability support can occur. The points in the figure are representative rather than exhaustive. Each design effort demands a different set of activities for the usability

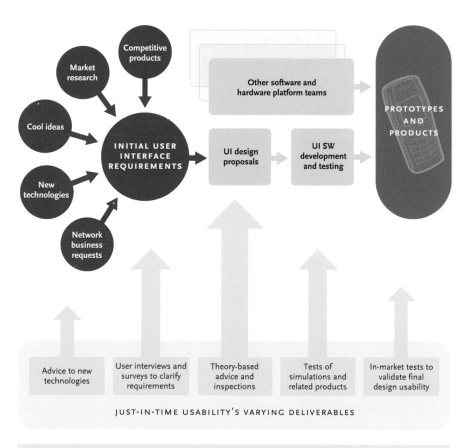

Figure 11.1 Just-in-time usability in the product creation process.

expert, depending on the scope of the change and the questions that surface during design. Usability activities are not planned by a central
authority, but are "pulled" from the usability expert as needed, just as
inventory items are pulled by *kanban* cards in the Japanese factory. To
the extent that planning is required, it, too, is pulled from the usability
expert. And each expert, like every other team member, has a personal
responsibility to envision future needs and be prepared to meet them on
demand.

This is the just-in-time approach. In Nokia's usability engineering, as
in the Japanese factory, the defining features are rapid and flexible
responses to novel situations, team responsibility for the quality and
timeliness of the output, the broad expertise of the individual team
members, and an opportunistic attitude that is especially effective in
supporting continuous improvement of the entire product line. From a
functionalist perspective, the similarity between the Nokia approach
and the Japanese method is not surprising. Both are rational responses
to the need for efficient use of resources in the face of rapidly changing,
hard-to-predict demands. Indeed, the founder of the just-in-time system, Toyota's Taiichi Ohno, suggests that Henry Ford might have
designed such a system if faced with today's business and technological
constraints.[5]

Just-in-time processes address the weak points of traditional planning
by defining a "lean production" system that is appropriate to the modern
business environment. The Japanese system deals with the relatively
well-defined issue of unpredictable demands for factory output. Nokia
responds to the more complex issue of unpredictable technological development and chaotic interactions among emerging systems. In both systems, the key is a Zen-like technique of applying just the right force at
just the right time.

Recall the Zen apprentice who stands in awe of the master butcher.
The master's knife never dulls, and his arm never tires, because he never
resorts to force. He carves by gently guiding the knife along its natural
course. In the same way, Nokia's design and usability teams forgo the
lock-step development routines that might characterize a more structured
company. The path unfolds, and we lead by following it.

References

1. S. Shingo, *A Study of the Toyota Production System from an Industrial Engineering Viewpoint* (rev. ed.). Cambridge, Mass.: Productivity Press, 1989.

2. Y. Monden, *Toyota Production System, an Integrated Approach to Just-In-Time*, 2d ed. Norcross, Ga.: Institute of Industrial Engineers, 1983.

3. J. P. Alston, *The American Samurai.* Berlin/New York: De Gruyter, 1986.

4. R. L. Harmon & L. D. Peterson, *Reinventing the Factory.* New York/London: The Free Press, 1990.

5. I. Majima, *The Shift to JIT, How People Make the Difference.* Cambridge, Mass.: Productivity Press, 1992.

6. M. J. Schniederjans, *Topics in Just-in-Time Management.* Boston: Allyn & Bacon, MA. 1993.

7. J. Gould and C. Lewis, "Designing for Usability: Key Principles and What Designers Think," *Commun. ACM* **28**(3): 300–311 (1985).

8. J. M. Carroll, ed., *Scenario-Based Design: Envisioning Work and Technology in System Development.* New York: Wiley, 1995.

9. T. K. Landauer, *The Trouble with Computers.* Cambridge, Mass.: MIT Press, 1995.

10. C. Lewis and J. Rieman, *Task-Centered User Interface Design: A Practical Introduction.* Shareware book available at *ftp.cs.colorado.edu/pub/cs/distribs/clewis/HCIDesign.*

11. J. Nielsen, "Guerrilla HCI: Using Discount Usability Engineering to Penetrate the Intimidation Barrier," in *Cost-Justifying Usability*, R. G. Bias and D. J. Mayhew, eds. San Francisco: Morgan Kaufmann, 1994, pp. 245–272.

12. D. Schuler and A. Namioka, eds., *Participatory Design.* Hillsdale, N.J.: Lawrence Erlbaum Associates, 1993.

13. H. Beyer and K. Holtzblatt, *Contextual Design.* San Francisco: Morgan Kaufmann,

14. J. D. Gould, S. J. Boies, S. Levy, J. T. Richards, and J. Schoonard, "The 1984 Olympic Message System: A Test of Behavioral Principles of System Design," *Commun. ACM* **30**(9): 758–769 (1987).

15. W. D. Gray, B. E. John, and M. E. Atwood, "Project Ernestine: Validating GOMS for Predicting and Explaining Real-World Task Performance," *Human Comput. Interaction* 8(3): 237–309 (1993).

16. S. Bødker, P. Ehn, J. Kammersgaard, M. Kyng, and Y. Sundblad, "A Utopian Experience," in *Computers and Democracy—a Scandinavian Challenge*, G. Bjerknes, P. Ehn, and M. Kyng, eds., Aldershot, UK: Avebury, 1987, pp. 251–278.

17. S. J. Grossman, G. Lynch, and M. Stempski, "Team Approach Improves User Interface for Instruments," *Electronic Design News*, June 4, 1992, pp. 129–134.

18. J. Nielsen and R. L. Mack, eds., *Usability Inspection Methods.* New York: Wiley, 1994.

19. C. Wharton, J. Rieman, C. Lewis, and P. Polson, "The Cognitive Walkthrough Method: A Practitioner's Guide," in *Usability Inspection Methods,* J. Nielsen and R. L. Mack, eds. New York: Wiley, 1994.

20. R. G. Bias, "The Pluralistic Usability Walkthrough: Coordinated Empathies," in *Usability Inspection Methods,* J. Nielsen and R. L. Mack, eds. New York: Wiley, 1984, pp. 63–76.

21. B. Hayes-Roth and F. Hayes-Roth, "A Cognitive Model of Planning," *Cognitive Sci.* **3**(21): 275–310 (1979).

PART 4

Mobile Contradictions

He does not stand on rock, but on shaky ground. Still it seems to hold his weight pretty well. Actually it is better this way: he can balance with a little effort and the environment keeps him alert. This is an excellent position for now, he thinks. Yet he knows he can't stay here long without losing his footing.

Far down the horizon there's still no sign of solid terrain, but he is not overly concerned. He believes there will eventually be a toehold if he keeps fighting his way forward. In spite of the haze, he is convinced he's heading in the right direction. What makes him really uneasy is the very next step. It is so close; it is so chancy.

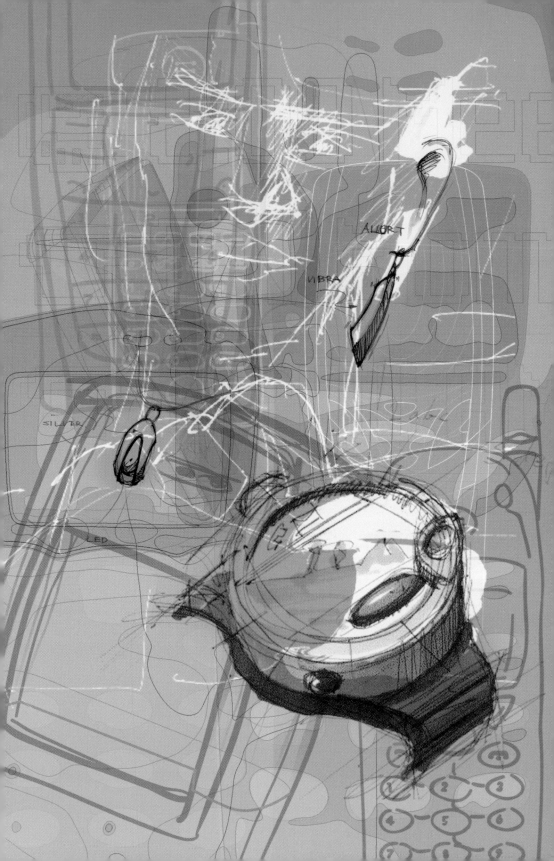

CHAPTER 12

The Expressive Interface

The consumer electronics industry has always had the biggest hills to climb when trying to make every product truly user-friendly. In many other industries millions are spent on designing and testing complex systems that are completely contained; the system is used only within a certain set of formal constraints, whether they are location-based, training-related, and so on. The operator of a spacecraft control computer (see Chap. 2) was initially selected for having a certain aptitude and was then trained for many months before being given the responsibility of actually using the computer. Even your average office PC user will have some rudimentary education on how to use the computer and office software applications.

In the world of consumer electronics, things are different:

- Anyone can buy your product—assuming that they can afford it.
- Few will take the time to read the complete manual—even when we run into problems, most of us will struggle with menus and workarounds for some time before admitting defeat.
- All consumers must get what they want out of the product in terms of value.

So far we have had reasonably simple consumer electronics products to design, such as mobile phones, TVs, or microwave ovens. With these devices, consumers have limited things to do—make a call, change the channel, or reheat yesterday's dinner. And they have only limited ways to do it—tap in a phone number, reach for the remote control, or set the cooking time.

In general we can say that the usability of our products is relatively good. Designing usability down to the lowest common denominator does make for a very usable—if occasionally dull—product. But if we work on the principle that each buyer or user is an individual (a point emphasized in Chaps. 4 and 5) and that the number of possible options buyers want is increasing exponentially, then we either find a better way or run into some very big problems.

A Menu with 200 Items

Smart products like mobile phones are increasingly made out of software. They are powered by general-purpose processors and supplied with several megabytes of memory. The displays are small, but resolution is improving and many manufacturers have introduced color screens. This sort of platform presents an opportunity to create a vast number of features without incurring any extra costs in hardware. The costs of software design and implementation are minor when divided by the number of similar products produced. This implies that whenever there is extra memory, developers will fill it with features. It would seem a waste of memory and processing power not to.

Market-driven pressures to increase the number of features in mobile phones are continual. By increasing their functionality, the products can be made to appeal to a larger market segment. The more features there are, the more likely it is that consumers will find some that they feel they need (see, e.g., the results by Pekkarinen and Salo in Chap. 10). Sometimes consumers can't try out a real product when purchasing a mobile phone. Perhaps they have access to a dummy model, or (more likely) can only view the models locked in a glass display case. In those cases, the length of the feature list is a strong sales argument.

How do those features manifest themselves to the user? In a mobile phone, all features are accessed with a small number of keys, some of which can be activated in different modes. Such keys change their behavior according to current phone status at the moment they are pressed. Typically, the features of a mobile phone are organized as a hierarchical menu structure (see Fig. 12.1 and Chap. 2).

Figure 12.1 3310 phone menu hierarchy illustration.

Let's make a highly simplified calculation of the average mobile phone feature set. The hierarchical menu structure of a mobile phone is like a tree; each function can be accessed from the root of the tree (main menu) through several branches (submenus) down to the desired function (leaf). The tree is called *balanced* if all the leaves are equally as many branches away from the root. If all the approximately 200 features of the Nokia 3310 phone were arranged in the menu as a balanced tree, we could minimize the average number of key presses that a user is required to make. Selecting the menu requires one key press (menu softkey), moving to the next menu item one key press (arrow down), and selection of either another submenu or a feature one key press (select softkey). With a balanced tree, the average number of key presses exceeds 12. A phone that requires users to make a dozen selections in order to complete a frequently undertaken task, like placing a call, is not a usable device.

A mobile product must be small to be carried and held in one hand with ease. There is simply not enough room for many more keys. Nor can key size be reduced much more because of the limits of human physiology. On the other hand, research by Keinonen[1] shows that the user's perception of a product's usability depends on the number of keys on the product; more keys equal a product that is harder to use. If consumers want a phone that is particularly easy to use, they will probably select the one that has the fewest buttons. This behavior can lead to disappointment—what looks easy to use may actually be complicated to operate because of the vast number of features it contains.

Some researchers have explored ways of minimizing the number of keys on a mobile phone. Harri Wikberg describes a user interface with just one key,[2] and some Asian manufacturers have sold single-key numeric pagers. Although these efforts are largely regarded as curiosities, it is still important to push the limits of how minimal user interfaces can be and yet retain a degree of ease of use.

Interaction designers of mobile phones face a challenging, almost paradoxical task. The number of features is rapidly increasing. At the same time, the number of user interface (UI) controls with which to access features is being reduced.

For Better or Worse

Designers struggled mightily to solve the problem of the menu with 200 items (i.e., the Nokia 3310 mobile phone), because they found that pressing 12 keys to reach each function wastes the user's time. In addition to the design challenge, this problem raised another question: Is speed important?

In the study of human–computer interaction, the quality of a user interface is called *usability*. Jakob Nielsen, a distinguished usability expert, defines the concept to consist of *learnability, memorability, efficiency, error rate, and satisfaction*.[3] His definitions and the heuristic rules that he proposes for spotting usability defects have been used extensively. The four first attributes are relatively easy to measure with a stopwatch (hours, minutes, seconds, milliseconds) and with counters (number of errors). These can be easily turned into value (dollars, euros, yens) by multiplying the wasted work time with the hourly rate of the unproductive time. The last attribute, satisfaction, is much more difficult to express in numbers. Naturally, the users can be asked to rate on a scale of 1 to 5 how much they agree with the sentence "I felt in control?" and several other claims. Although user satisfaction can be converted to a numerical value in principle, it is close to impossible to convert a satisfaction rating into a monetary value.

Traditionally, usability specialists have concentrated on making systems efficient and accurate. The satisfaction attribute has been used for evoking the users' subjective evaluation of efficiency and accuracy. Those evaluations ignored the fact that there are many inefficient, relatively inaccurate systems with which users are extremely satisfied. Consider the following examples:

- *Games.* These are intentionally difficult to learn and master. Three errors typically mean that the user needs to start the game all over again.
- *Virtual chatrooms.* Users spend time wandering around a 3D world before talking to strangers in the virtual chatroom. They could achieve the same amount of chat much faster if it were text-based.
- *Websites.* Many Websites use unnecessary graphics, GIFs, flash animations, and other features. Content could be delivered more efficiently with plain text.
- *Mobile phones.* When users are changing color covers, downloading icons and ringing tones, and playing games with their phone, do they really achieve something we could call *efficiency?*
- *Electronic banking.* The users regularly go through an unnecessary number of steps because the bank wants users to feel that their transactions are safe and confidential.

In addition, designers of personal technology such as mobile phones and handheld computers must think of the product as a companion to the users. People may spend more time in a relationship with a device than with any of their friends. As in all human relationships, attitudes toward an electronic companion change. In the beginning, the fun of exploring the possibilities of a device and the sheer pride of ownership can engender very positive feelings about a product. Later, when the user gets to know all the product features and perhaps finds some unexpected disappointments, these feelings can change to boredom or dissatisfaction.

It is no longer sufficient to design products that support users' tasks efficiently and without undue errors. Personal technology must also fulfill emotional needs, which are much more difficult to evaluate and design for than the practical ones. These products need to live with each user in a longlasting relationship, avoiding premature divorce.

Spoiled by Choice

Imagine a happy couch potato lounging around with a large box of popcorn, a couple of beers, and a wide screen TV connected to cable and satellite. He may find himself in something of a dilemma. Having a choice of 500 channels, how does he select the program that he wants to watch? A traditional paper-based *TV Guide* is far too unwieldy and probably out of date. So he has three choices:

- *To watch what he normally watches.* It's 7:45 p.m., so that means "Friends."
- *To go channel hopping.* If we assume 3 seconds per channel, in this case it will take 25 minutes to surf through them all. Supposing that the

Scenario sketches drawn in a project aiming at concepts for short distance wireless communication. *(Keinonen 2000.)*

program he decides to watch was somewhere around channel 450, he will have missed the show, anyway.

- *To use the electronic program guide.* Even on a wide-screen TV, this display can only show a small percentage of what is available. Therefore the guide can be made intelligent enough to support a search feature—this couch potato wants to watch an action film. The latest electronic program guides have progressed to suggesting what viewers might like to watch on the basis of their TV watching profile, whether determined automatically or based on a questionnaire.

This degree of complexity is just about to hit mobile phones. The person in the street will soon have something in common with the overburdened TV viewer, except that the screen of this individual's handheld unit is considerably smaller. We can already say with some assurance that 200 items in the hierarchical menu of the Nokia 3310 is too much for easy and efficient navigation. Still, the phone doesn't even include a browser of any kind—WAP or Web—as its successor Nokia 3330 does. Let's do another quick exercise. What will happen when all the information in the Internet is available through mobile phones? Typing the search term "mobile phone" in AltaVista (*www.altavista.com*) results in 727,822 hits [June 20, 2001 at around 15:00 GMT (3:00 p.m. Greenwich mean time)]. If we were to perform the search on a mobile phone, the browser would display 727,822 items in one list or in a hierarchical tree to the user. At that point, designers would face a problem of a new magnitude. The hierarchical menu structure that has been appropriate so far would collapse. Would it be simply impossible to access a large amount of information, such as those the Internet delivers, through a small mobile device?

The Most Important Person Is Ann

Most mobile phones have a memory for storing names and phone numbers. The phonebook application can be easily accessed with a single key press. The phonebook itself consists of up to 200 or 300 names in alphabetical order. Finding a desired name in the phonebook will—as in the menu—mean approximately 12 key presses to reach a listing and to make a call. If a person's name happens to be in the beginning of the alphabetical order, for example, "Ann Adams," it can be reached very quickly. On the other hand, poor Margaret Mayer is much more difficult to reach.

Most certainly, for any mobile phone user, not all names in the phone book are equal. If we ask randomly selected persons in the street what the most important aspects of their lives are, they typically include other people in the top three answers. In addition to good health and happiness, they will say "my wife," "my boyfriend," "my family," or "good friends." Although they always refer to one person or a very small group of people that matter most, the mobile phonebook always facilitates communication with names high up in the alphabetical order.*

This simple observation that the significance order of contacts is not the same as the alphabetical order of the phonebook—or any other rigid mechanical way of arranging contacts—is a start in recognizing that the mobile phones are not merely technical objects storing a fixed number of entries. They have the potential to be representations of the most valuable things in the user's life: relationships with other people. Mobile phones could be designed in such a way that they would support creating, enjoying, and maintaining these relationships. The phone could provide features that help the user stay in contact with loved ones all the time. It could transmit information about the caller's current usage context

---

*Many users fool the automatic alphabetic order by adding a prefix to the important people, for example, writing "AAA—Johanna" instead of "Johanna."

to others. Users could know what their friends are doing, where they are, who they are with, and how they feel.

Mixed Emotions

It has been said that the Navi-key user interface sometimes "feels magical" (see Chap. 3); that is, it seems to guess what the user wants to do next. Careful analysis of user needs and probable subsequent actions results in a design that is efficient to use, but it also evokes an emotional reaction of surprise and admiration. Have the designers intentionally created a product that causes positive emotional reaction?

Emotions are very difficult to analyze. They are people's immediate and subconscious reaction to changes in the environment, and beyond that nobody has been able to properly define what emotions actually are. There's no definite and widely accepted list of all possible emotions. Even if there were, emotions are usually mixed with each other: joy with surprise; anger with sorrow; or possibly more than two at the same time. Even as the conscious mind tries to decipher the emotion, it changes or passes.

To design for a desired emotional effect, we would first have to measure the emotional effect of a product. There are currently two possible ways to attempt it. First, it is possible to measure physiological phenomena that occur when a person is experiencing emotions. These include facial expressions and changes in the person's voice, galvanic skin response, and electrocardiographic, electroencephalographic, and electrooculographic (EKG, EEG, EOG) tracings. The challenge in this form of analysis ("affective computing") is that there is no straightforward formula to calculate the emotions based on the measurements.[4]

The other approach is to use subjective analysis. Users of the product are asked about their feelings before, during, or after use. The challenge of subjective analysis is that it is problematic for most people to identify their own emotions precisely. How questions are worded in any attitude scale (there are several in use) can easily influence the rating because emotions are difficult to describe. We've even seen some attempts to avoid the problems inherent in expressing emotions through language by using pictures and animation to describe these emotions.[5] Although in its infancy, the research has the potential of providing important information about specific rather than general emotions. It is naturally more interesting to the designers to know if the users feel fascination or pride rather than generally finding the product pleasant.

Given the intractability of these obstacles, designers can take a third approach: design by intuition and empathy. Designing for emotional effect doesn't necessarily have to be measurable to the finest granularity. Good results can be obtained simply by allowing emotions to be a design driver in product design. In this case the designer becomes a sort of surrogate for the user. The designer can use different stimuli to put the user in the right mood for designing emotional products and can also collect information from the end users that contain emotional clues.[6]

Let's say that a mobile phone designer wants to experiment with emotional design, and decides to create a mobile phone that will not get too predictable—or boring—during the long period of product ownership. The phone currently provides a standard feedback notification "message sent" when a text message has been successfully launched. The same feedback could be provided to the user in hundreds of different ways. The feedback might be an animation and a subtle sound effect that is different every time. The user could never expect the subtle feedback that the phone is providing this time. Still, the usability would not be compromised. When all the animations and sounds have been used, the phone could automatically download new ones from the network. Similarly, all other features of the current phone could be redesigned to provide a surprising effect.

Some emotional designs may require the deployment of advanced technologies. Consider the "funny" phone. Right now devices are static; they contain only the features and information that have been programmed in them. A device that tells the same joke twice is not funny. To continue being funny, it will therefore need to change its behavior. This requires that the device adapt to the user. Using affective computing technology, the device could actually sense what kind of jokes the user typically laughs at and present only jokes of that type. Naturally, the device can log which jokes the user has already seen to avoid presenting the same joke twice. When devices can be connected to the Internet and their users form a community, we can imagine an ecosystem where each device borrows material from the interactions of other users with their *similar* devices. Such devices could learn hundreds of new jokes *every* day for an indefinite period of time.

The Machine Knows What You Need

Eventually the vast amount of information on the Internet will be available to users through generic browsers in mobile phones. Is there some way we can help users find information useful to them among all the hundreds of thousands of links? We have already determined that designers should not approach this problem by designing a static unbalanced hierarchical tree, as with mobile phone features, because the number and relevance of items in a search result are not known at design time. In order to minimize the interaction paths that conduct users from queries to the answers they need, the system must dynamically create an unbalanced tree as the search begins.

Presumably that can be done. Mobile phone features are prioritized on the basis of a user needs analysis. The most important and most frequently used features should be the easiest to access. With information searches, some search results are likewise more significant than others, and some search engines leverage that fact using traditional information search methods. All Webpages that contain the search string are first listed as search results. Then the pages are analyzed more closely to discover, for example, how many times the search string is mentioned in a given document. More advanced text analysis tools also understand the context in which the searched term appears in the text.

These methods are wholly dependent on the information in documents turned up by the search and the search terms provided. Because mobile phones are companions that remain with their users throughout the day, they have a much wider potential for refining search results. As users access information through their mobile phones, the phone can keep logs of the kinds of information in which the users are often interested. It can track the users' communication patterns—the places they visit and the persons to whom they send messages. By reference to the calendar, the phone knows something about recent and future events to which the search results might be related.

Let's play the scenario out into the future. A user's phone activities probably have patterns dependent on the time of day. In the morning, local weather and traffic information about the user's route to work are predictably interesting. An advanced phone could be equipped with sensors to analyze the environment and determine where the phone is. Positioning capabilities are already emerging. The desired traffic information is naturally going to change depending on whether the user is leaving home or leaving the office.

If a phone is possessed of context awareness and can adapt to user behavior, then it can use the same information to change user interface components such as the menu hierarchy. Elementary forms of such adaptation already exist. For instance, most phones have a dynamic list containing the most recently dialed numbers. This means that the system has logged user behavior and provided a way of easily repeating recent actions. Currently call lists are ordered by calling chronology only. The most recently dialed number is first on the list, the number dialed before that is second, and so on. Here, one simple way of introducing more intelligence would be to add other ways to sort the list. It might be worthwhile to rank the most frequently called numbers. As the users probably call different numbers at different times of day, the order should change itself by dynamic. During the working day, the most frequently called persons are usually colleagues and customers. On the drive home from the office, users probably call home, and in the evening they call friends and relatives. In informal studies we have found that the five most frequently called numbers *weighed by the time of day* constitute 75 percent of all calls.

The possibility of adaptations like the ones we've just imagined presents yet another design challenge—when should adaptations occur? Does the user interface adapt and change gradually, increment by increment, in an evolutionary manner that might go unnoticed by the user? Should it be more decisive, changing big time at infrequent intervals, thereby surprising the hell out of the end user for a short period of time and then becoming consistent again? Of course, both of these suggestions could be wrong—the first is just too slow for a product that may be replaced every 6 to 24 months, and the second stands a good chance of putting users off the product for life.

Discussions on adaptivity for personalization have so far focused on the positive benefits that it can bring to bear on overcoming the increased complexity of the phone and mushrooming amounts of information that it can deliver. On the other hand, not all products or even all aspects of a product need to be adaptive. Knowing the difference is key. Adaptivity

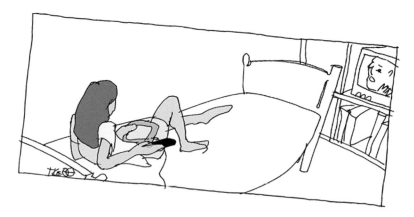

must not be used to fix poor interaction design, and should not be used to bemuse. It adds value to the product when it increases the user's odds of successfully completing a task.

Bringing information about the user and context into the design equation has risks. It requires that a device store a lot of personal information about the user—an amount way beyond the user's comprehension. What would happen if someone else got hold of this information? What security issues need to be addressed to avoid misuse of personal information, to create trust, and to encourage the user to grasp the opportunities of adaptivity? All in all, it is agreed that user interfaces could be greatly improved by using all the information that mobile devices acquire about users and their environments. This will simply have to be done judiciously.

The Permissible Intrusion

Are we sometimes too pious when it comes to security and privacy? When the conversation turns to privacy, in the industry everyone's first thoughts (to avoid being sued) are

Privacy is valuable
Privacy must be guarded
We can't allow any intrusion on rights of privacy

All this is very true; privacy is a big issue. But instead of thinking about compromises to privacy only in terms of threats, could they also be seen as

an opportunity? What if users would be willing to give up some forms of privacy in a controlled manner? Are there situations in which they might decide to do that? And what kind of gain would make it worthwhile?

One ready example is retail store "loyalty" cards. Shoppers who use loyalty cards to get a small discount are also allowing their shopping preferences to be logged into the store's databases. In this case, users don't seem overly self-protective. The benefit for them is very straightforward: a monetary discount. They sell this sort of private information for a relatively cheap price.

Another kind of gain that users may get in exchange for private information is access to another person's private information. Here the motivations are less straightforward. What value can be realized from this sort of swap? In the case of complete strangers, not much, but if the people involved are family members or close friends, the information could be quite precious. What personal information would you like to share with your nearest and dearest right now? You probably wouldn't mind letting your children know where you are and what you're doing. You might be happy to let your spouse know, for instance, whether anything special is happening, if you're feeling better, or if this is a good time to call. Close friends could get almost as much information as family. Colleagues could have access to all your work-related information, such as whether you are at your desk. Will you be available for a quick chat before leaving the office? Acquaintances and associates may have access to very little of your private information—maybe the equivalent of what's on your business card. In addition, you could elect to make certain information public, like a published Webpage, and available to everyone.

In each of these scenarios, access to the user's private information is, and must be, fully under the user's control. As a default, private information would be available to nobody. But since technology-mediated information exists apart from the users, these users must also be able to trust device manufacturers, service providers, and network operators that private information is used only for the purposes that they have permitted.

Currently the only way I can share personal information with you is to tell you. I can call you or send an email. In Europe the short message services are used extensively for this purpose. The typical content of a text message is not factual, like the time of the next meeting or flight reserva-

tion details, but emotional: greetings, jokes, expressions of feelings, say-ing "good night" or "I love you." If such information were available all the time, the intended recipient could access emotional content without any action by the sender. This would increase the sharing of truly per-sonal information, enhancing and intensifying the relationships between people.

The mobile phone would be an excellent platform for sharing private information. It is carried on the user's person most of the day. It automati-cally knows the user's communication patterns. The phone even pos-sesses information about the use context. For example, the user already can put her phone in a meeting mode to prevent it from ringing aloud. The phone already has a calendar application that contains information about where, and with whom, the user is at the moment. For a set of trusted persons, all this information could be mediated by a personal device, much as it once was by a private secretary.

User's Dream—Designer's Nightmare

In an idealized version of the future, mobile phones will be designed care-fully to meet our needs. The experience of using mobile phones will be both efficient and emotionally satisfying. The devices will learn our habits, and use that knowledge to organize features and content in the order most effective and relevant to us. Our closest friends will have controlled access to our private information. This enhances and intensifies the relationships with our loved ones, family, friends, and colleagues. We will get more attached to our mobile phones because the phones reflect the things that we find important. The phones will become true mobile companions to us.

The important themes of the future—good interaction design, emotional expressiveness, adaptivity, sharing, and privacy—are interestingly entangled. Good interaction design has to be combined with an understanding of the emotional value of the product. To overcome information overload, adaptive products can log user activity to improve information search accuracy. Adaptation and user customization are different techniques for the same goal. The accumulation of information about the user in adaptive systems is a privacy risk. On the other hand, if we want to create products with emotional value, the device needs to know the user to respond in a meaningful way. Information itself is a privacy risk, but it becomes emotionally significant only when shared with other people.

Because of the integral links between these themes, designers won't have the luxury of considering them one at a time when building future products. It is an attractive idea to include them all in a futuristic product concept. The result would, however, probably be unacceptable to users, who would have too many abstract principles to adopt. These themes are more properly introduced to the market slowly, in small steps at a time, and without hype. For the user, the future seems interesting, if confusing. For the designer, despite all the challenges, the future confers potential to change and enhance the way people live.

References

1. T. Keinonen, *One-Dimensional Usability—Influence of Usability on Consumer's Product Preference,* PhD dissertation, University of Industrial Arts, Helsinki, 1998.

2. H. Wikberg, "How Small Can We Go? A Mobile Phone User Interface Based on Single Key," in *OzCHI 2000 Conference Proceedings,* Sydney, Dec. 2000.

3. J. Nielsen, *Usability Engineering.* San Diego: Academic Press, 1993.

4. R. Picard, *Affective Computing.* Cambridge, Mass.: MIT Press, 1997.

5. P. Desmet, P. Hekkert, and J. Jacobs, "When a Car Makes You Smile: Development and Application of an Instrument to Measure Product Emotions," *Adv. Consumer Research* **27:** 111–117 (2000).

6. B. Gaver, T. Dunne, and E. Pacenti, "Design: Cultural Probes," *ACM Interactions* **6**(1): 21–29 (1999).

6 mobile messages received

read

Turkka Keinonen and Christian Lindholm

CHAPTER 13

Six Mobile Messages

We have written here about mobile user interface development at Nokia. We describe the role and elements of the fundamental UI building block, the user interface style, in mobile handsets. We illustrate what happened when we set out to design a phone for second-generation digital mobile communication technologies in the 1990s. We discuss different methodologies and approaches to studying end users, carrying out usability research, and conducting usability engineering activities within the constraints of daily product development work. We also ponder the landscape ahead of us, as it seems from our point of view in the haze. We have based what we say on our own experiences. There have been no comparisons to our competitors. Still, we believe that many of our experiences and approaches are not company-specific, but characterize the whole field of mobile communications and perhaps even user interface design for any portable interactive device. Where our thoughts differ, they can be used as a benchmark.

One thing we have not done is to present a single principle behind what we do. Instead, we've tried to illustrate that user interface design is not like the huge boulder in the lobby of the corporate headquarters, but more like the gravel it would become if it were crushed and spread throughout the research and product creation divisions.

Some overall principles, though, are recognizable. They are related to the very basic constraints that we work with: mobility, consumer markets, and flexible and rapidly developing technologies. In this chapter we will consolidate these six mobile messages for your convenience.

Small User Interfaces Do Not Scale

Mobile applications need to be redesigned for different terminals, because a small user interface does not scale down. Feature prioritization is crucial.

The mobility of users—their mobile interaction with the portable terminal and their consumption of information and entertainment in mobile situations—makes a difference in user interface design. A mobile user interface is not a miniaturized desktop UI.

In particular, the small user interface does not scale down. When adding features and squeezing them into the device, developers face a usability "knee" (Fig. 1.5). That's the situation where the UI starts to become too packed; in principle, new features can be accommodated, but in practice they would not be very usable any more. There is a limit in the density of a user interface; a menu structure, a screen resolution or an icon of given size has an inherently limited human-machine communication bandwidth.

Applications require a minimum screen space to be usable. When the screen size decreases, one cannot simply keep compacting the presentation, but instead the whole structure of the user interface must be reconceived. In a calendar application, the concepts of a monthly or weekly view make no sense if the screen does not allow for displaying the dates and a certain amount of user-defined information about the entries. At that point the metaphor needs to be changed from "calendar" to "list of reminders," for instance. Applications have to be optimized for small screens on all layers of the interface: the vocabulary layer, the logic layer, and the layer of functionality.

A causal relationship obtains between portability, user needs, and the details of the UI. *Portability* implies requirements concerning, for example, size, one-hand operability, and usability in multiple task situations. These specifics influence the selection of user interface hardware components and input solutions. With limited screen size and key count, any presentation of alternative options has to be sequential; only a few items can be presented at a time. Prioritization then becomes crucial, as the most-needed options have to be among those presented first.

In every situation, creating approachable and easy-to-use interfaces depends on nothing so much as the priority of functions. In the case of basic phones, the top-priority user action is often close to self-evident— there may be just one thing that's really necessary while all the rest are secondary options. One key may be enough. On the other end of the complexity continuum, mobile terminals are used to access the same content as desktop environments with full-size keyboards, large displays, and broadband connections to the Internet.

Fragmented Minds—Do Not Standardize

Users are cognitive, emotional, contextual and cultural actors. It takes segmentation, personalization, and continuous evolution to fulfill their versatile changing needs

Excerpts from Focus Group Study Results: Study Conducted in October 2001

Ambiguous responses to family communication concepts. Eighteen respondents (6 fathers, 6 mothers, 6 teenage girls) were interviewed in three focus group sessions about five concepts for mobile services to support communication within families.

New technologies rejected—new features required. Respondents' perspectives on new solutions were confused. Even their basic attitude toward the concepts included acceptance and rejection side by side. On one hand, we heard disparagement of the "technological imperative" more than once. On the other hand, some of our service concepts were rejected because they didn't exhibit enough genuine new benefits as compared to existing solutions. . . . Mobile, terminal-independent access is not necessarily a benefit as such if the consumer doesn't see the value of it.

> When you do sports, you do sports. No extra devices needed. I'm very much against the idea that our leisure time should be made so efficient. If we are "in chains" ourselves, do we have to chain our children, too?

> I don't like the idea that technology takes the responsibility from adolescents. They should learn to take responsibility themselves.

> It doesn't matter if we miss one training session with my son because of a last-minute change. Actually, that would give us a good chance to spend some shared time in peace for an hour.

Limited interests to share beyond a core group of intimates. People value their private time. Interest in close contact with others is limited to one's own family and a few good friends. Parents don't have much time and energy to sustain a wide range of social contacts. Making a point of keeping up with distant relatives was considered "mental," not commendable. What motivates social community is shared interests and mutual goals, not just being a relative or neighbor. Parents were experienced users of information technology (IT) at work. They had experienced frustrations about using IT in that context. To them, leisure time is for privacy and relaxation, which leaves little room for new technologies. Face-to-face contact is preferred if you have something emotionally loaded to say.

> One of my friends got invited to a funeral by SMS, and someone told me about a friend's divorce by SMS. In my opinion that is completely tasteless and unacceptable.

Insulting associations to undesirable groups. Associating product features and usage scenarios to the wrong groups can be insulting. Finnish mothers got really upset when our designs seemed to connect them with household issues and children. Some working mothers said that they are so bombarded by that aspect of life that they'd prefer hobby-related scenarios instead. Associations with secondary comprehensive school implied by the scenarios insulted sixth-form girls.

> After three years of staying awake nights with wailing infants, you'd rather see some stories about women having fun with friends.

Utility discussion rules. Utility discussion dominates the focus groups. This may be partly a methodological bias, but clearly it is also due to communication habits. Adults say that serious issues should be handled face to face or by voice calls. New fun and pleasure-optimized features are regarded almost like toys, and that's a negative association. Toys are for kids, not for adults. Teenagers consider themselves adults or want to associate themselves with adults. Consequently, the objective for leisure communication is that the solutions have to be serious communication tools that can also be used for fun.

. . . on selected occasions—with 100% user control. Continuous communication of . . . needs to be under deliberate control. . . . Thus, the system needs to support . . . only for controlled specific reasons, in specific situations.

Only phone features are accepted. New . . . solutions need to be presented as features of a mobile phone. Operator service solutions, such as WAP services, caused a lot of mistrust. The reason was not completely clear, but probably it was related to expectations of costs, complexity, and difficulty of operation. When features are realized as network services, they must be seamlessly integrated with the terminal's UI. New device categories for special purposes were not accepted.

The market for mobile phones is a consumer product market. The easy approachability of user interfaces is an obvious objective, but not the whole objective. A person's relationship with an artifact looks different depending on the angle of observation, and several angles may be relevant.

Every now and then I'm correctly interpreted as an information-processing unit accomplishing practical tasks with my handset. The next time I may be an actor trying to adapt my behavior to the surrounding cir-cumstances. In other words, what I do makes sense only when seen in specific context. I may be a consumer choosing handsets, using handsets, exhibiting handsets, and speaking about handsets to construct and sup-port my lifestyle and identity. I may be a participant in the wider cultural discourse evaluating mobile phones by socially constructed meanings. My behavior may be driven by my utilitarian or hedonistic motivations. I am a member of a culture, shaped and constrained by it. I'm also chang-ing all the time along with the world around me, and occasionally I even do something myself to initiate a change.

Simple user segmentation models such as the ones based on demo-graphic information or frequency of use are not enough. Individuals differ in several respects, even within themselves. Basic task-oriented usability is not enough. To improve user interface designers' understanding of the multidimensional consumer, the designer must be empowered to see the shifting relevance of design tasks and to focus on what is essential.

Comprehensive user understanding has become topical for several reasons. A business expands into market areas and cultures that are rela-tively poorly known compared to the company's previous experience. Simultaneously, in those markets where mobile technology has been most widely accepted, it has become an essential aspect of social life. Availing oneself of mobile phones and services is losing its novelty value. The enthusiasm of the user's first foray has continued for a fairly long time, and there is reason to get ready for normalized markets. Naive users are being replaced by hard-to-satisfy people who are accustomed to versatile IT applications and devices. Usage and consumption motivations and pat-terns have to be classified better than before.

The fragmentation of the user population has two kinds of implica-tions: those for the method set and those for the product portfolio. Longi-tudinal studies, contextual design, and cultural research are tools intended to deepen our insight into the consumers' lives. The tools come from sociological and ethnographic research traditions, and are being adjusted for the product development environment.

One size does not fit all. A range of models is needed, and product segmentation has to percolate to the level of user interfaces: screen sizes, control elements, operational logic, features, language, shortcuts, graphics, and sounds. All the components of user interaction need to be designed so as to satisfy versatile demands. Simple design assumptions, such as using English as the default UI language in the design phase and then translating the vocabulary later, just do not work. Other languages have essentially longer expressions to be accommodated on the small screens. The layout design needs to be verified with several languages.

Providing UI diversity starts with the user interface style portfolio and ends in UI personalization.

Strive for a Seamless User Experience

The mobile industry faces a wireless complexity threshold. To overcome it we will have to provide a seamless user experience of terminals, applications and services.

Enhancing Your Daily Mobile Communication

The world is moving toward universal connectivity. . . . X gives you a new kind of freedom and convenient connection without . . . requirements. X and . . . are designed for people who desire freedom and flexibility in data communication.

Acknowledgment: www.nokia.com.

Summary Report (Company Confidential)—Phase 2 [Product X]

LESS COMPLEXITY A MUST FOR X TO ADD VALUE

Twenty-four respondents had been using X intensively for 4 weeks by June 2001, when they were interviewed in depth about their experiences with the product.

RECOMMENDATIONS

Apart from solving problems with reliability, the following improvements are needed to make X valuable to the business user:

- *One-click connection between phone and* . . . X adds a layer of hassle and complexity which made many respondents prefer to go back to using . . . [present solution] rather than X. Managing multiple X . . . should be offered as an option only to very advanced users. The X software should be invisible to the average user.

- *PC X should mirror the phone UI.* Many respondents discovered a whole new side of their phone when having used X . . . In particular . . . suddenly became a highly valuable function on the 6210 . . . However, PC X should more closely mirror the phone; its UI should behave like a phone UI to avoid learning barriers, and the functionality of the PC X should correspond more closely to all the things that you can do on the phone.

- *Avoid repeating the "WAP is crap" problem.* Less knowledgeable business users had expectations that X would cure all ills: that it would give faster connections to the Internet, that it would allow them to. . . . Expectations need to be managed, and the technology is better marketed on its specific application benefits than on the general X name.

  Respondents in this research were mostly advanced mobile data users, but they nevertheless had problems using X and getting value from the product. Since X is a pioneering product, they were forgiving about its flaws. However, they were expecting a more mature product from the company they trust to deliver easy-to-use solutions.

Mobile phones continue a tradition of easy-to-use household products. A telephone is a known quantity; everybody understands its purpose and its basic operation. The first step toward mobile technologies consisted of nothing more than removing the wire. Now, content, applications, and services are exploding. Digital convergence, connectivity, mobile services, downloadable applications, mobile multimedia, location-based services, SyncML, Bluetooth, Wireless Village, and other mobile technology buzzwords lead to complex products and complicated applications. Good usability will become more difficult to provide. Design flaws will occur. The discipline is facing a wireless complexity threshold resulting from user interface density, off-line configurations, and usability incompatibility.

User interfaces for the second generation of communication products, namely, GSM terminals, were designed based on a set of functions that were considered sufficient and mature for implementation in the mid-1990s. Now the same user interfaces are being stretched thin with new features. If the UI solutions are flexible enough, these novelties can be realized in a manner that is reasonably usable as long as we look at each one in isolation. However, the amount of new features in itself adds additional layers to the user interface. There will be new menus, new settings, and longer lists of options. The right ones will be increasingly difficult to find. The overall complexity increases, and the old interfaces grow dense.

Many of the new systems therefore aim at automated services and a seamless user experience. In basic use, once all knobs and dials are set where they're supposed to be, there should be no problems. But then there is the matter of configuring services. The services—and their usage scenarios—are unknown to the phone manufacturers, who naturally find it difficult to define the appropriate system behavior. Thus they prefer to leave the decision to the end user. A priori offline settings are typically more difficult for the user to manage than real-time control, where the response from the system is immediately perceptible.

The mobile terminals, applications, services, and sometimes even service content originate from different sources. They may be technically compatible, but technical compatibility is different from "usability compatibility." A wireless service may be optimized for a terminal—or technology—completely different from yours. You'll need to reconfigure the device to get connection, and perhaps you'll even have to reverse the new configuration when you go back to the normal-usage modes. By contrast, features that are embedded in the product can be presented in an intuitive manner in the user interface. The user needs only to browse through the menus and check the available options. When a service depends on the cooperation of several devices and multiple communication technologies, all the options cannot be presented to the user. Often the user needs to know in advance what to look for—and where to look for it.

The wireless complexity threshold is impervious to any amount of usability work that is focused on services, applications, or terminals alone. We need to focus on producing seamless user experiences, where all the different layers of technology and service provisioning are streamlined to create end-user comfort and value.

Think by Doing

> For mobiles, fancy visions and real end-user value conflict. Taking the next step requires the resolve to make instead of dream and decide instead of speculate.

The importance of products fulfilling real user needs is repeated over and over again, but the mobile world around us is not constructed on practical needs alone. Solutions can be explained after their success to fulfill some need when, very likely, without the introduction of the designs themselves, the depth or even the existence of those needs would have never manifested itself. Consider the contemporary mobile content services as an example—what is selling is downloadable ringing tones and screen logos. The practical utility of these services is close to nonexistent, yet consumers use them, and some people have made their living producing them. Without having seen this firsthand, few of us would have believed that such a business could exist.

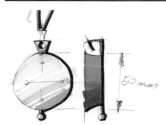

Hand-drawn sketch
Project team:
"This is cool. I'd like to have one. The users will love it."
Users:
"This is boring."

PC simulation I
The review board:
"We want something more Nokia-like."

PC simulation II
Every designer:
"Oh no, The beautiful simplicity of the initial concept is gone"
Project manager:
"Let's try to add more fun to the previous one. Could you try to make games without a display."

PC simulation III
Designer:
"I like this, the users probably won't."
Project manager:
"Nah, we can't do decent games without a display."

Hardware prototype to the board:
"That's too close to basic phones. Let's kill the project and make a phone feature out of your concept."

Future mobile services promise wild possibilities. They are not yet concrete, but they point the way in promising directions. There is a strong will and commitment to a mobile information society. However, taking the next step to introduce a new service that is truly discontinuous with present technology is the headache. It would be very convenient to launch it based on a set of hard-core practical needs. The examples in this book about mobile service studies underline this point; personally relevant services that save time and trouble are a reliable basis on which to design new technologies.

The foreground of mobile interfaces is likewise full of possibilities: adaptability and intelligent user interfaces, virtual communities and sharing, emotional interaction, and all other emerging technologies. The vision is there, but what is the first adaptable feature that really adds value to the end user?

The Nokia user interface development culture appreciates down-to-earth thinking. In many business units at Nokia, aversion to hype is almost tangible. Fancy, exploratory projects without links to known technology changes or user needs have been few and far between. We hope that examples in Part 3 have given readers some idea about the concrete and near-term objectives of the preponderance of projects. They reflect the passion for doing instead of dreaming, for business thinking over academic thinking, and for deciding over explaining.

Research and concept design aim at the straightforward concretization of ideas. Much of the thinking takes place through doing. A range of prototyping techniques and research approaches are in use. Different kinds of problems are approached by designing and evaluating different UI simulations: e.g., paper prototypes for menu navigation and hardware simulations for text entry and dual-task problems. Research and design methods are seldom applied by the book. The real objective is to get something to work. If we learn why it works, that is a by-product—and an extremely welcome one at that.

The "rough and dirty" approaches have been mentioned a couple of times in this book. Some of those projects have turned out to be relatively large and time-consuming, however. The notion of "rough and dirty" needs to be understood as referring to the attitude of focusing on the essential rather than actually running through projects as superficially as possible. Arranging global tests, prolonged testing with prototypes for learning curves, arranging dual-task situations, and creating mobile services all require a lot of work and expertise. Subjects may tolerate prototypes for only a short while.

There is also a level of usability decision making which occurs when there is no time for user studies. Innovative technologies come from somewhere else than user needs in such cases, but there may still be valid reasons for incorporating them into products. The usability experts need to work for the best possible match in every circumstance, no matter how averse. Ultimately the organization must trust in the expertise of the team and the competence of its individual members.

"Doing," that is, working with prototypes, becomes increasingly challenging. Novel services require field testing. The equipment must be functional, portable, and suitable for field use. Several prototypes must be manufactured so that they can communicate realistically. Packing technology into a fieldproof form at the prototype phase is substantially more difficult than building simulations intended for demonstration and laboratory test use.

The More You Polish the Better It Gets

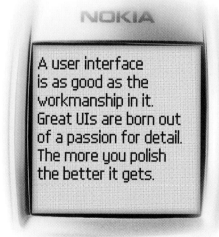

>>>———Original Message———
>>>From: NN (NMP-RD/Espoo)
>>>Sent: 23. May 2000 15:38

>>>To:

>>>Subject: Right alignment of digits

>>>

>>>Hello Super Gurus,

>>>

>>>Why are we aligning the phone numbers to the right when everything
>>>else is on the left? I was thinking that you must know the reason. Is the
>>>right alignment a Nokia heritage holy cow, that must be maintained no
>>>matter what? Otherwise, are there any real reasons to keep it right
>>>aligned (except that changing from right to left would mean more work
>>>for SW)?

>>>

>>>Background information:

>>>

>>>The right alignment may cause some problems when we start adding
>>>more functionality to the idle display. . . . The functionality will overlap
>>>with digit entry for call making.

>>>

>>>BR,

>>>NN

>>——Original Message——

>>From: NN (NMP/Salo)

>>Sent: 24 May, 2000 14:52

>>To: . . .

>>Subject: RE: Right alignment of digits

>>

>>Hi all,

>>

>>my understanding of the history (for a more authoritative one we should
>>ask NN):

>>

>>In the Good Old Days phone displays were damn small and could only fit
>>a few characters, it was hard to do any layout design at all, but right
>>alignment was probably one way to improve the user experience, because:
>>-when the number was so long that all of it could not fit on the display

>>the software would naturally drop the first digits, which were considered
>>less important for the user. (For names, the beginning is more important).
>>-It is also a way to make a visual difference between number and name
>>modes.
>>
>>And I do agree that there is no real reason to keep the tradition alive any
>>more.
>>
>>BR,
>> -N
>——Original Message——
>From: NN (NMP/Copenhagen)
>Sent: 24 May, 2000 16:26
>To: . . .
>Subject: RE: Right alignment of digits
>
>Hi there,
>
>Good points, N
>My 'guess' would have been:
>
>-Heritage at that time (1991 or so) was probably mainly derived from
>calculators (and NMT phones, where the digits come from right :-)
>-Most competitors did it that way back then (and they still do)
>
>I didn't quite get how right alignment conflicts with your concept.
>
>BRN
——Original Message——
From: NN (NMP/Salo)
Sent: 25. May 2000 10:55
To: . . .
Subject: RE: Right alignment of digits

Hi Gents,

Before we do anything else, let's see a simulation or test software with the proposed change. Maybe it's a case of "not broken, don't fix it." And we have plenty of genuine UI problems right now.

Benefits to making the change might be enhanced if we had some automatic number grouping (which is, I know, never perfect, but which might help the one true annoyance, which is that it is hard to check a very long number without losing your place).

If you want history, letting numbers come from a different corner from Alphas gives users a subtle, subconscious clue to whether he's in a number or alpha entry mode. Your eye knows exactly where to find the digit that you just entered. This is a bit different psychologically from alpha entry. In alpha entry you see where you are by "reading" from left to right, the whole thing each time. In number entry, if you do NOT have nice short number grouping, you tend to check and confirm one digit at a time. And the lower right position for the number echoes the format of the name and number in the phone book display, and how numbers are entered in calculator mode. It does not echo date or time entry.

I love these detailed esoteric UI discussions. These details are what have made our UIs the best. I hope someday we actually start spending time on them again! Meanwhile . . . is full of small besser-wisser irritations which were originally corrected 10 years ago (such as stupid extra notes everywhere, or that horrible beep when battery is full!!!!). Let's fix those first please!

User interface design is about constructing links from the big picture to the smallest detail. The wireless communication technology itself is layered. There are the capabilities of GSM, CDMA, GPRS, WCDMA, and other mobile protocols that are utilized within the limits of memory, display, and the capabilities of different terminals as specified by the user interface design. There are the mobile content providers and operators who are creating a selection of mobile services that appear to the end user to be specified by the user interface design. There is the user in a cultural context with more or less practical needs, which are being fulfilled to the extent specified by the user interface design. There are several chains of reasoning that lead, at the end of the day, to pixel-level UI solutions. Understanding where the chains come from is important, but still, the devil is in the details.

It is never easy to see the big picture, but sometimes it seems easier than designing the details. However, that's probably because there is help available for the big picture. We can consult marketing research, business analysis, and roadmapping to illuminate future trends. When it comes to the details, though, the full responsibility is on user interface designers and usability specialists.

Seen from the level of a product vision, problems with the details may seem trivial—it is, after all, just a matter of packing things into space, time, and hierarchies. However, the user interface is as good as the details in it, and the product is as good as the user interface.

Design for Development Stability

The challenge of mobile UI development is shared among external software firms, industry consortiums, and service providers. Reasonable development stability is a must.

The Open Mobile Alliance aims to grow the market for the entire mobile industry by enabling subscribers to use interoperable mobile services across markets, operators, and mobile terminals. This is achieved by defining an open-standards-based framework to permit applications and services to be built, deployed, and managed efficiently and reliably in a multivendor environment.

The objectives of the Open Mobile Alliance are to:

- Enable consumer access to interoperable and easy-to-use mobile services across geographies, operators, and mobile terminals
- Define an open-standards-based framework to permit services to be built, deployed, and managed efficiently and reliably in a multivendor environment
- Establish one mobile industry standards forum, the Open Mobile Alliance, to function as the driving force responsible for creating service level interoperability
- Drive the implementation of open services and interface standards, through the user-centric approach to ensure the fast, wide adoption of mobile services. www.openmobilealliance.org

The responsibility for developing mobile communication user interfaces is spreading. The user interface previously belonged to the individual terminal manufacturer, with influence from the mobile operators. Today the main stakeholders in the industry form consortia to ensure compatibility of technologies over different terminals and networks. Interfaces are being opened up to third-party developers. Developers and mobile service providers need a stable environment in which to propose new features and services. They also want the environment to be as widely used as possible to make their business profitable.

Mobile phone user interface designers cannot simply ignore these changes in the business environment and continue to pioneer new interface approaches every time they encounter new ideas. Proprietary solutions lacking technical compatibility with the overall business value chain will be very difficult to sell. The leading companies have both the ability and the responsibility to influence the direction of user interface development, but the direction must be clear and foreseeable to the rest of the industry. The terminals and networks alone without rich and usable service offerings will not satisfy customers.

Acknowledgments

The editors wish to acknowledge Erik Anderson, Seppo Helle, Harri Kiljander, Anne Kirjavainen, Christian Kraft, and Harri Wikberg.

TELECOMMUNICATION GLOSSARY

2G, 3G   G stands for mobile communication generations, i.e., 3G refers to the third generation of mobile communication technologies. The first generation of mobile technology, such as Advanced Mobile Phone Service (AMPS), uses nonstandardized analog radio systems. 2G systems, such as Global System for Mobile communications (GSM) and Personal Communications Services (PCS), use digital radio technology for improved quality and a broader range of services. 3G is a set of digital technologies that promises improvements in capacity, speed, and efficiency. Users of 3G devices and networks are promised an access to the kind of multimedia services envisioned only by science fiction writers, such as video-on-demand, video conferencing, fast web access, and file transfer. Finland holds pride of place in 3G, having been granted the first UMTS licenses in early 1999. Data transmission speed is one key improvement 3G has over its predecessors. Some say 3G networks will provide transmission of data up to 2 megabits per second (Mbps), although others assert that current network capacities will likely slow this to around 384 Kbps. Still, even the slower speeds are quite an improvement over current 2G networks, which transfer data at just 9.6 Kbps.

Bluetooth   Bluetooth technology is a low-power radio technology being developed to deliver short-range wireless mobility. Bluetooth technology holds the potential of eliminating intraoffice cables between devices, such as computers and printers, and providing services when devices are in close proximity. Bluetooth technology is an open standard and operates in the 2.4 Gigahertz (Ghz) frequency band. Devices utilizing Bluetooth technology should be able to transfer data—digital information—at transmission speeds up to 720 kilobits per second (Kbps). While this is too slow for the kind of media-rich experience we are getting used to on the Web, it is more than sufficient for transmission of the relatively short packets of data required. Bluetooth began development in 1998 under the auspices of the Bluetooth Special Interest Group (Bluetooth SIG), a group comprised of computer and telecommunications companies, which now number some 1400 member institutions, including 3Com, Ericsson, IBM, Intel, Lucent, Microsoft, Motorola, Nokia, and Toshiba.

CDMA Code Division Multiple Access is a technique in which the radio transmissions using the same frequency band are coded in a way that a signal from a certain transmitter can be received only by certain receivers. CDMA is based on what's called spread spectrum technology. The technique was first pioneered during World War II, as a method used for hiding communications and to prevent jamming by the enemy. CDMA was first used for civilian communications in the 1980s. As a cellular technology, CDMA was specified by the Telecommunications Industry Association (TIA), and is referred to as IS-95. CDMA uses both digital and analog techniques, which allow multiple users to occupy the same frequency without interference. Today, some 47 million users worldwide use CDMA networks.

GPRS General Packet Radio Service is a standard for wireless communication whose transmission speed is 150 Kilobits per second (Kbps). GPRS is particularly good for mobile Internet applications such as sending and receiving email. GPRS also offers instant access and permanent connection between the mobile device and the network. GPRS promises to enable a wide range of mobile applications now stunted by the slow transmission speeds of current wireless networks. GPRS is an important stepping-stone toward third generation (3G) networks. There have been GPRS terminals on the market starting from 2001.

GSM Global System for Mobile Communications was originally a European digital system for mobile communications. It was first introduced in 1991. Now GSM has become the de facto standard in many regions around the world, serving more than 100 nations. The notable exception is the United States, where adoption of GSM is still in its infancy, and analog networks still dominate. More than 239 million people around the world use GSM networks. Technologically, GSM uses what is known as narrowband Time Division Multiple Access (TDMA), which allows eight simultaneous calls on the same frequency. GSM works primarily in three frequencies:

GSM 900, GSM 1900, and GSM 1800. The GSM 900 system is the most extensively used worldwide. GSM 1900 is primarily used in urban areas in the United States. GSM 1800 is primarily used in urban areas in Europe.

HSCSD High Speed Circuit Switched Data is an upgrade to GSM networks that enables data rates to increase to 57.6 Kbps. HSCSD was introduced in 1999 to upgrade the GSM data rate from the previous maximum of 14.4 Kbps.

ITU The International Telecommunication Union is the international organization within which governments and private companies coordinate telecom networks and services.

MMS Multimedia Messaging Service is a new standard that is being defined for use in advanced wireless terminals. The service allows for non-real-time transmission of various kinds of multimedia contents like images, audio, video clips, etc.

NMT Nordic Mobile Telephone is an analog cellular system originally developed by Ericsson for use in Finland, Sweden, Denmark, Norway, and Iceland. NMT is operated in 450- and 900-MHz bands.

PDC Personal Digital Communication is a digital system for mobile communications in Japan.

SIM SIM stands for Subscriber Identification Module, a module that is inserted into a GSM mobile device for subscriber identification and other security related information.

SMS Short Message Service is service used in mobile communication systems by which a user can send or receive short messages—up to 160 characters—in textual form. SMS, as it is generally known, has become widely popular in Europe and the Far East since 1997, although the technology has been around since 1992. As of October of 1998, 2 billion short messages were being sent per month on GSM networks. Most SMS messages are sent person-to-person as simple text (e.g., "Meet me at the bar, 17:30"), but it also supports mobile information services, such as news, sports, stocks, weather, horoscopes, SMS chat, notifications, and downloadable ring-tones and icons.

SynchML SynchML is an open synchronization platform, which recognizes the need for a single data synchronization protocol—

supported by over 500 service providers, application developers and manufacturers of mobile phones and other communication devices. This will make it possible to get up-to-date information with any application.

TDMA Time division multiple access (TDMA) is digital transmission technology that allows a number of users to access a single radio-frequency (RF) channel without interference by allocating unique time slots to each user within each channel. The TDMA digital transmission scheme multiplexes three signals over a single channel. The current TDMA standard for cellular divides a single channel into six time slots, with each signal using two slots, providing a 3 to 1 gain in capacity over advanced mobile-phone service (AMPS). Each caller is assigned a specific time slot for transmission. Because of its adoption by the European standard GSM, the Japanese Digital Cellular (JDC), and North American Digital Cellular (NADC), TDMA and its variants are currently the technology of choice throughout the world.

UMTS Universal Mobile Telecommunications System is a third generation (3G) mobile communications technology that promises data transmission speeds of up to 2 megabits per second (Mbps), although actual speeds may be significantly lower at first, due to network capacity restrictions. It is expected to become commercially available in Europe in 2003.

W-CDMA Short for wideband CDMA, a high-speed 3G mobile wireless technology with the capacity to offer higher data speeds than CDMA. WCDMA can reach speeds of up to 2 Mbps for voice, video, data, and image transmission. WCDMA was adopted as a standard by the ITU under the name "IMT-2000 direct spread."

WAP Wireless Application Protocol is an open global standard for mobile solutions, including communication between a mobile device and the Internet or other computer applications. Because of the standard's support in the industry and its compatibility with many existing standards (GSM, CDMA, TDMA, as well as with emerging 3G standards), WAP is expected to meet wide-ranging acceptance with consumers. Even now, mobile operators, like Vodafone and EuroTel, and Internet portals, like msn.com and Yahoo!, are racing to build WAP services. While the current version of WAP doesn't support the kind of media-rich content people are used to getting from the Web, it does allow specially tailored Web-based content and services to be used on WAP mobile phones and devices. WAP facilitates such services as stock trading, mobile banking, and airline reservations. The WAP specification was announced in June of 1999 by the WAP Forum, which is an industry organization that brings together companies from all segments of the wireless sector.

INDEX